Haynes

Bee Manual

First published in June 2011
Reprinted 2012 and 2013

British Library Cataloguing in Publication Data
A catalogue record for this book is available from the British Library

ISBN 978 0 85733 057 4

Library of Congress control no. 2010939708

Published by Haynes Publishing,
Sparkford, Yeovil, Somerset BA22 7JJ, UK
Tel: 01963 442030 Fax: 01963 440001
Int. tel: +44 1963 442030 Int. fax: +44 1963 440001
E-mail: sales@haynes.co.uk
Website: www.haynes.co.uk

Haynes North America Inc.
861 Lawrence Drive, Newbury Park,
California 91320, USA

Printed and bound in the USA by Odcombe Press LP,
1299 Bridgestone Parkway, La Vergne, TN 37086

Author's Acknowledgements
The authors would like to thank Dr Gay Marris of the National Bee Unit, Fera, Paul and Gill Smith of EH Thorne (Beehives) Ltd, Dr Kathy Keatley Garvey of the University of California and Bill Stevens of National Bee Supplies for their assistance during the preparation of this Manual.

Credits

Authors:	**Claire and Adrian Waring**
Project Manager:	**Louise McIntyre**
Copy editor:	**Ian Heath**
Page design:	**Richard Parsons**
Index:	**Carol Ball**
Illustrations:	**Dominic Stickland**
Photography:	**Claire Waring (unless stated otherwise).**

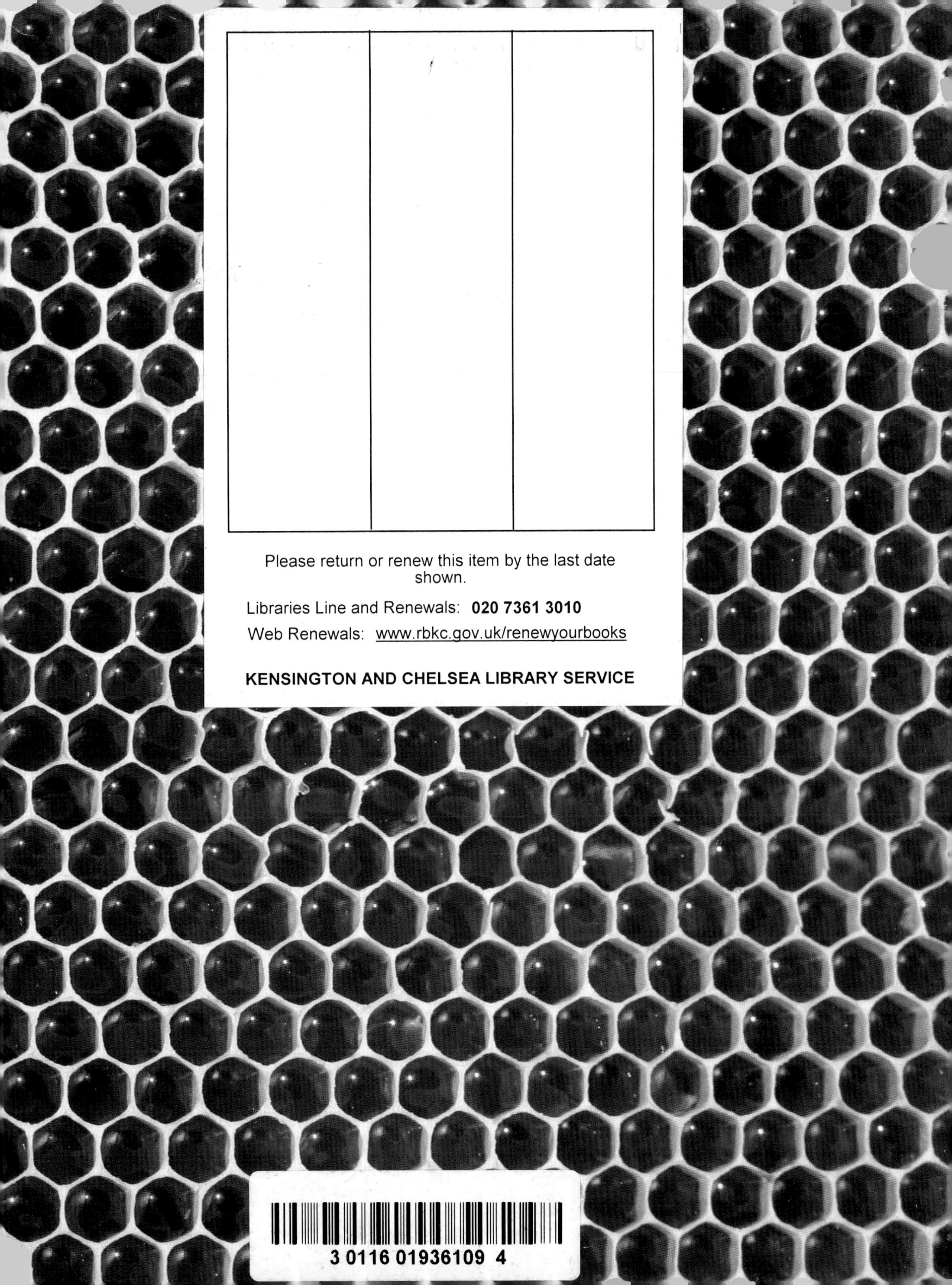

Haynes

Bee Manual

The complete step-by-step guide to keeping bees

Claire & Adrian Waring

Foreword by Bill Turnbull

CONTENTS

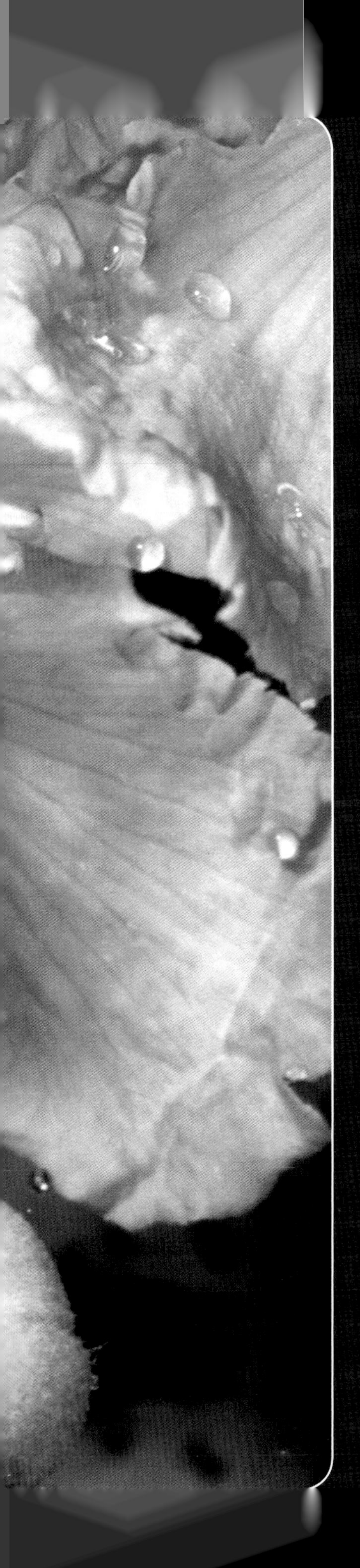

FOREWORD

If there's one thing I know about beekeeping, it's that you can never have too many books on the subject. Over the past ten years, I've collected a small library of them. A couple are more than 100 years old. One, Eva Crane's *World History of Beekeeping and Honey Hunting*, is now so rare that it cost me more than £100.

But amid the veritable forest of beekeeping tomes, where should you start? This manual by Claire and Adrian Waring is an excellent way to begin.

There are an awful lot of different ways to keep bees. And an awful lot more mistakes you can make in the attempt. Believe me, I should know – I've made loads of them. If only I'd had access to a Haynes Manual at the time I'd have understood more easily the difference between top and bottom 'bee space' (very important) or how to tell between slotting the frames 'warm way' and 'cold way' (also quite important).

What's really useful here is the illustrations. Beekeeping equipment can be quite hard to describe. Did you, for instance, have any idea what a mouse-guard looks like? Here, there are loads of pictures offering an accurate representation of what the authors are talking about. Six whole photos and a diagram to explain the process of lighting the smoker! That's the beekeeping equivalent of learning to ride a bike; not easily learned, but never forgotten.

In fact, if you read this book and practise what it teaches, you could probably take up beekeeping completely by yourself. But I wouldn't recommend that – and nor do the authors by the way. So many things can go wrong, and in my case over the past ten years almost all of them have. It helps to have an experienced hand around to guide you back onto the straight and narrow; so it's important to join a local association if you can. If, on the other hand, you are reading this on a desert island with just bees for company, then this is the best companion you could have to get you going.

As for the authors, you don't get much more expert than Claire and Adrian. They've both been keeping bees for yonks. They've both been General Secretary of the British Beekeepers' Association. They've written a string of other books on the art of apiculture. In addition, Adrian is a former county beekeeping instructor, and Claire is editor of *Bee Craft* – the monthly magazine that is a 'must-read' for British beekeepers. In short, they know what they're talking about.

Beekeeping is a fascinating, challenging and highly rewarding pastime. Harvesting the honey is just a small part of it. What I love is the rest of the process: learning to live with your bees, enjoying the privilege of delving into the hive, and above all keeping them alive. It's not always easy. Part of the beekeeper's lot is learning to live with the setbacks. But with this book, Claire and Adrian are at your shoulder. So you're in good company.

Good luck!

Bill Turnbull

Introduction

I wonder why you became interested in bees and beekeeping? Recently, the plight of the honey bee and other insect pollinators has been highlighted in the media and this has certainly made many more people aware of these fascinating insects. It's now known that thousands of honey bee colonies have been lost in the USA to a phenomenon called Colony Collapse Disorder (CCD). We have also become aware that colony deaths in the UK have been higher than normal over the past few years. Attention has also been drawn to the sharp reduction in the amount of suitable forage available to pollinators.

This situation has stimulated great support for bees and beekeeping. The Women's Institute passed a resolution at its 2009 Annual General Meeting entitled 'SOS for Honey Bees', calling on the government to increase funding for research into bee health. Support for beekeeping has also come from corporate bodies such as The Co-operative and Rowse Honey, and all round the country gardeners are being encouraged to plant

Above: Other pollinators include bumblebees but these are not kept by beekeepers.

Below: The honey bee, *Apis mellifera*.

bee-friendly flowers. The government has allocated additional money for bee research and announced a joint initiative with other partners to research and identify threats to bees and other pollinators.

However, perhaps you're like me. I have always been interested in the natural world and I remember going out on my bicycle with my father to collect wild flowers when I was young. In those days wild flowers were not legally protected as they are now, and we only ever took one specimen for identification purposes. Looking back, my specific interest in bees possibly stems from my primary school days. I can still picture our geography teacher catching a swarm and hiving it in a white WBC hive on top of the air-raid shelter. However, it was not until much later that this interest was rekindled and I actually decided I wanted to keep bees.

Thus started an amazing hobby. Firstly there's the thrill of having your own colony of bees and watching them all around the hive. Then there's the scary moment when you do your first colony inspection on your own, and the panic when you realise your bees are preparing to swarm and you haven't finished making the hive you want to put them in. I doubt any beekeeper forgets the excitement of watching their very first honey crop flowing out of the extractor. I even still have one of my very first jars of honey!

Talking to beekeepers, I find many of them had contact with bees when they were young. For some it was their parents, grandparents or other relatives who kept bees. For others it was someone coming into school to talk about bees and maybe set up an observation hive which was then used for projects in science, maths, art, English and other subjects. That's one of the nice things about bees – they can have a bearing on so many aspects of life.

Whatever has stimulated your interest, I should warn you that you're very likely to get hooked. After a couple of years you may begin to think that you've learnt it all and everything will be plain sailing. However, talk to any experienced beekeeper and you'll find that this is definitely not the case. With bees, you're always learning.

As you become more and more absorbed by your interest in your bees, you may well find yourself extending that interest in different directions. You may start selecting and breeding your own queens. You may begin collecting bee books. Then there is bee photography, which brings you into the macro world. Another possible extension is into the study of other insects, such as solitary bees and bumblebees. You may even find yourself becoming fascinated by other bee species found round the world, opening up the opportunity to travel.

This book cannot hope to cover all of these aspects, but it is designed to give you a clear introduction both to bees and to the practical aspects of beekeeping.

I hope you enjoy it and that you and your bees will benefit from it.

Above: Watching bees in an observation hive is always fascinating.

Below: Beekeeping can be for all ages.

THE BEES

Why keep bees?

Man has utilised or kept bees in one way or another for centuries. Initially, our main interest was in their honey – the only sweetener available until beet and cane sugar were refined – and beeswax, which could be burnt to produce light, used as a carrier for medicinal herbs and oils, and as a polish. Man would, and still does in some parts of the world, rob honey from bees' nests. If the honey was squeezed out of the comb rather than the whole thing being eaten, the remaining beeswax would then be melted, filtered and utilised.

However, while the honey and beeswax harvest remain an important part of beekeeping, this aspect is far outweighed by the importance of bees in crop pollination. In the UK, honey bees are estimated to be worth £825 million to the agricultural economy. It's said that every third mouthful of food we eat is available because of the honey bee. What a staggering thought! And it's not just food for humans – bees also pollinate crops for seed, such as carrots and onions.

In fact, the honey bee is the third most important domestic animal after cattle and pigs and ranks ahead of poultry.

Above: A honey bee on oilseed rape.

Below: As well as producing honey, bees perform a vital pollination service.

Above: Learning about bees.

Man and honey bees

So, man needs the honey bee, which is the only source of honey and beeswax. On the other hand, the honey bee now needs man, since the arrival in the UK of the *Varroa destructor* mite in 1992. If left untreated an infected colony will die, and this has led to the demise of virtually all feral honey bee colonies in the United Kingdom.

Away from the altruistic idea that the beekeeper is the saviour of the bee, there are many other reasons for keeping bees. You'll be dealing with a fascinating insect that is probably the most closely researched in the world. It's an insect that communicates the distance and direction of a food source by dancing on the surface of a vertical honeycomb in the complete darkness of the hive. It's an insect that lives in a colony which can be described as a superorganism, with each individual contributing in some way to the survival of all. And it's an insect which can see in colour but which switches to black and white while flying at speed to conserve resources.

The more you delve into the life and behaviour of the bee, the more you'll want to know. As a beekeeper you'll need to learn to 'read' a colony of bees to determine the situation – whether they're about to swarm; whether they need more room to store honey; whether they're healthy or need treatment.

You'll learn about swarming and swarm control and you'll join many others who are surprised by their bees' actions and declare that they don't read the books. You'll begin to appreciate different temperaments in different colonies and learn how to deal with this. You'll see your bees visiting flowers in your garden and experience the thrill of taking your own honey harvest.

Above all, you'll have the joy of watching and handling your bees and you'll have fun!

Beekeeping isn't for everyone

Beekeeping may not be for you, however passionate you are about nature. People have been known to come top of the class in the theory lessons and yet be totally unprepared and horrified when standing by an open hive containing a large colony of bees. For this reason, if for none other, I recommend you attend a series of beginners' lectures before paying out for bees, hives and other equipment. Such courses are run by many of the local county beekeeping associations across the country.

My second recommendation is that you join your local beekeeping association. Here you'll not only be able to learn from visiting lecturers and attend apiary meetings to learn about opening and handling a bee colony, you'll also find a group of like-minded people who will be only too willing to give help and advice. As a beginner, you're very likely to need this support.

You're already following my third recommendation, which is to read a good beekeeping book!

Below: Join a local beekeeping course.

How does a bee colony work?

Above: Honey bees on the comb.

The first time you look at a living colony of bees you'll see thousands of individuals, apparently acting at random. However, while individual bees do spend a lot of their time 'resting' (doing nothing apart from helping to maintain the colony's temperature), they are, in fact, all working together for the survival of the colony.

A honey bee colony can be regarded as a superorganism. Technically, this is defined as 'a collection of agents which can act in concert to produce phenomena governed by the collective', where 'phenomena' are any activities that the superorganism requires, such as collecting and storing food or searching for a new nest site. Individual bees cannot survive for long on their own. It's the combined activities of those individuals in a colony that enables it to function.

Beeswax comb

The basic structure on which a colony lives is comb, made from beeswax secreted from four pairs of glands on the underside of the workers' abdomens. The tiny wax scales produced are passed forward to the jaws, or mandibles, where they're manipulated and softened before being added to the comb.

Worker bees hang together in festoons when making wax, and start building comb at the upper surface of their nest cavity. It's thought that the comb structure evolved from circular cells placed next to each other. Because beeswax is pliable, where the cells abutted the walls flattened out, giving the familiar hexagonal shape. This is one of the most efficient structures in terms of use of materials and it's also structurally strong.

Hexagonal cells are built either side of a midrib, with two parallel sides vertical and the other two pairs coming to a point at top and bottom. The lower inside edge of the cell slopes slightly so that the opening is higher than the base. Again using the most efficient construction, cells on opposite sides of the midrib are offset so that they fit together, and you can see this if you hold an empty newly built comb up to the light.

Below: Secreting scales of beeswax.

Below: Beeswax scales on the floor of the hive.

Below: The offset hexagonal cells in beeswax comb.

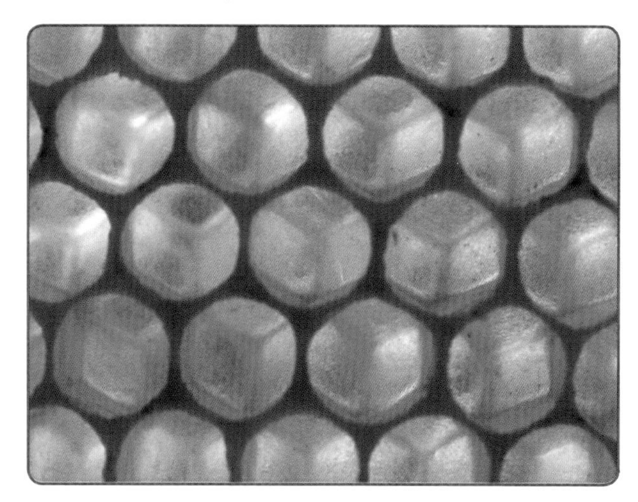

Cavity-nesting honey bees

Our honey bees – the Western honey bee, *Apis mellifera* – nest in cavities. They build parallel combs and the whole nest is protected by the cavity itself, with guard bees defending the entrance from predators. This is also true of the Eastern honey bee, *Apis cerana*. However, other honey bee species, such as the giant honey bees, *Apis dorsata* and *Apis laboriosa*, and the little honey bees, *Apis florea* and *Apis andreniformis*, make their nests in the open, suspended from a cliff overhang or a tree branch. They build a single comb that's protected from the elements and predators by layers of bees.

Above: Colonies of the giant honey bee, *Apis laboriosa*.

Below: Colonies of the little honey bee, *Apis andreniformis*.

Above: The queen.

The queen

A colony has one queen, which is the only bee able to lay fertilised eggs. In some circumstances the female worker bees may also lay eggs, but as they cannot mate these are unfertilised. Bee genetics is complicated, but you only need to understand that the queen can lay both fertilised and unfertilised eggs. Both are fertile, *ie* they'll hatch and develop into adults, but fertilised eggs develop into female workers while unfertilised eggs produce male drones.

A queen can live for four to five years, using the sperm she received when she mated. Her task within the superorganism is to produce the next generation of workers and drones. She also ensures colony cohesion as the pheromones (chemical signals) she produces are passed around the colony as workers share food.

Mating with the drone

The main role of the drone is to mate with a virgin queen, produced in the colony as a result of swarming, supersedure, or the death or loss of the old queen. Virgin queens and drones from neighbouring colonies fly to drone congregation areas (DCAs).
These are well-defined areas that are determined by the topography of the landscape. At the DCA, drones will investigate any object that might be a virgin queen, including stones thrown into the air by curious beekeepers. They form a 'comet', flying behind the queen, and then mate in the air. The act of

Below: Drones attracted to a feather impregnated with queen pheromone.

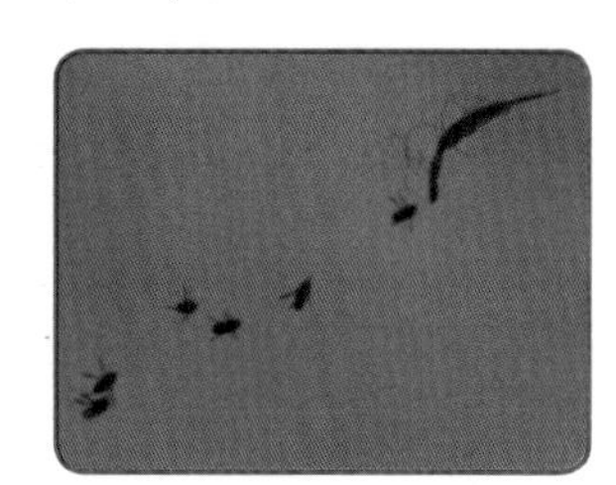

Above: The male drone.

Above: The female worker.

mating kills the drone, which falls away, and his place is then taken by the next one. We now know that virgin queens can mate with 10–20 drones.

The queen stores the sperm she receives in a small organ in her abdomen known as the spermatheca. This has a valve on the opening which allows her to fertilise an egg before she lays it, or not, as the case may be.

Drones cannot collect nectar or pollen and their survival depends on the provisions brought back to the nest by the workers. At the end of the active season when the colony is preparing to cluster for the winter, drones are a drain on resources and they are forcibly evicted or deliberately killed by the workers.

Above: A worker ejects a drone from the colony.

The worker

The vast majority of bees in a colony are female workers, which are responsible for most of the tasks required for colony survival. After a worker emerges from her cell she usually progresses through a number of activities within the colony as a 'house bee, or 'nurse bee', before joining the others collecting nectar, pollen, water and propolis (see below) as a forager or flying bee. However, research has shown that this progression is not fixed and bees of all ages can perform most tasks if this is necessary to meet colony needs. Their individual lifespans are linked to their nutritional intake rather than their energy expenditure.

The house bee starts by cleaning cells, ready for the queen to lay eggs in them. She then moves on to feeding larvae, first the older ones and then, as the hypopharyngeal or brood food glands in her head develop, the younger ones. She tends and feeds the queen and then, when her wax glands become active, she secretes beeswax and uses it to make or repair comb. At around 10–12 days old she starts receiving nectar from incoming foragers. The action of enzymes, coupled with a reduction in the water content, converts nectar to honey which is then stored in cells within the nest. The worker also packs down the pollen loads deposited loosely in the cells by foragers.

At about three weeks of age she's ready to become a forager. But before she takes up this role she may spend some time on guard duty, defending the entrance against robber bees from other colonies or pests such as wasps.

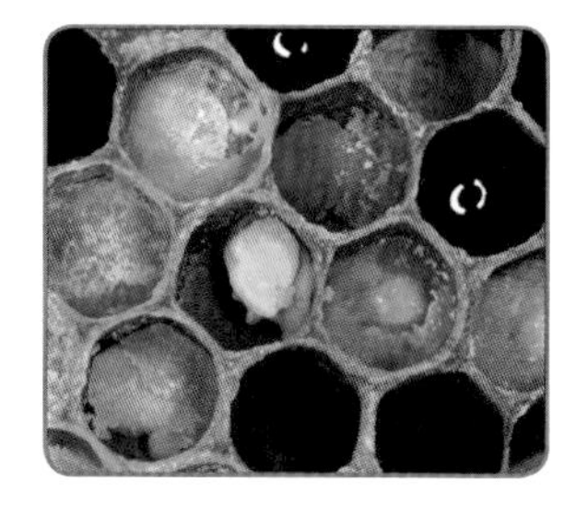

Above: Larvae (above) Pollen stores (below).

Below: A guard bee.

Above: Workers learning the position of their hive.

Above: A worker drinking.

Before she starts foraging the worker needs to learn the location of her nest. First she takes hovering flights facing the hive entrance. She then makes wider and wider circles until she feels confident enough to fly away from the hive.

Foraging

The worker's main foraging target is nectar, which provides the colony's source of carbohydrate, and she brings this back in her honey crop, situated within her thorax. She'll also bring back pollen, packed into the pollen baskets, or corbiculae, on her back legs. This provides the colony's protein, and when there are large numbers of larvae to be fed more pollen will be collected. Foragers will pick up pollen on their hairy bodies when collecting nectar, but they'll also forage specifically for it when needed.

The colony also needs water and foragers bring this back in their crops. Bees need water for their own survival, but they also use it in the nest to dilute stored honey before feeding it to the larvae, and to spread on the surface of the comb to cool the nest by evaporation.

Propolis is a sticky resin exuded by various trees, such as poplars. Foragers collect this, pack it into their pollen baskets and take it back to the nest. Here a house bee will help to unload it and then use it to fill up cracks in the hive or strengthen the surface edge of the combs. Some colonies use propolis to control the size of the entrance, reducing it to a few bee-sized holes in a curtain of propolis which otherwise covers the whole entrance.

Above: Propolis being taken into the hive.

Foraging is hard work and within two or three weeks the worker is worn out and will die. At the height of the active season she therefore lives only for some five to six weeks. However, when the colony clusters during the winter the worker's physiology alters and she's able to live five to six months. It's this, coupled with the bees' ability to store honey and pollen, that enables a colony to survive from year to year.

Below: A worker with a large pollen load.

Below: An old worker with wings frayed from use.

The individual bee

Bees are insects belonging to the Hymenoptera order. Their body is in three parts: head, thorax and abdomen. Three pairs of legs and two pairs of wings are all attached to the thorax. Each bee has two compound eyes and three simple eyes on the head, together with two antennae.

The male is the drone. However, there are two types (or castes) of female: the queen and the worker. Rather than describe the bee's anatomy in detail, it's probably more helpful to look at specific differences that are related to the tasks of the different forms in support of the colony.

The queen

The queen is the mother of all the bees in the colony. She has longer legs and a longer, more pointed abdomen than the worker. Many people expect her to be much larger than the workers but this is not so. The colony reproduces by swarming and the queen must be able to fly with the first or 'prime' swarm, so she's not that much bigger.

Above: A queen with her court of workers.

The drone

The drone is also slightly larger and much more burly than the worker. He can be distinguished by the square end to his abdomen and his very large compound eyes. All honey bees have five eyes – three 'simple' eyes placed on the top of the head and two large compound eyes on either side. These consist of many smaller eyes, each with its own nerve leading to the brain. In the drone, the compound eyes meet at the top of his head, pushing the simple eyes forward. They give him an exceptionally wide angle of view, which he needs when he flies to a drone congregation area to mate with a virgin queen.

Above: The burly drone.

The worker

The worker is the smallest and most numerous unit in the colony. Her main distinguishing feature is the pair of pollen baskets (corbiculae) on her hind legs. The leg joint is flattened and surrounded by a fringe of stiff hairs. Rows of stiff hairs on the inside of each leg are used to brush pollen grains from her body and pass them to the back legs. As she rubs her back legs together, the pollen is forced through

Below: A worker.

Above: Collecting pollen and taking it back to the hive.

a specially adapted joint into the pollen baskets. She takes the loads back to the nest where the pollen is fed to the larvae or stored for future use.

Most people are aware that bees can sting. However, they do not realise that the drone has no sting and that of the queen is smoother and used only during fights with rivals in the nest when more than one queen is present. This is generally during the swarming process, when several virgin queens are reared and they fight until only one remains, which flies out to mate before returning to head the colony.

The worker uses her sting to defend the colony. It is barbed, which allows it to hook into the enemy so that venom can be delivered at the sting site. However, the barbs prevent her from withdrawing it from a surface such as human skin, and when she flies off the sting mechanism tears from her body, resulting in her ultimate death.

The small eyes in the compound eye are each made from several cells, each of which is sensitive to light vibrating in a separate plane, enabling it to detect polarised light. Thus even if a bee can only see a little patch of blue sky it will know the position of the sun, and this is used to navigate to food sources communicated to foragers by round and waggle dances in the darkness of the hive.

Above: The sting.

In her head, the worker has 'brood food' or hypopharyngeal glands, which produce the substance fed to larvae. On the underside of her abdomen are four pairs of wax glands from which she secretes small particles of beeswax that are then passed to her mouthparts (mandibles) and moulded into shape for comb construction or cell capping.

Sexes and castes

There are also internal differences between the sexes and the castes.

The queen's larger abdomen contains two ovaries, where her eggs develop, and a spermatheca, which is a small vessel into which the sperm received during mating migrate. A valve at the entrance to the spermatheca allows the queen to release sperm to fertilise a worker egg or to withhold it to produce a non-fertilised egg, resulting in a drone.

Similarly, the drone's abdomen contains the testes that produce sperm. Because he needs to fly to the drone congregation area to mate he has strong flight muscles in his broad thorax.

While the worker has external pollen baskets to carry pollen back to the nest, she has a crop or honey stomach in her thorax that she uses to carry nectar. This is sucked from the flower with her tongue and regurgitated when she returns to the colony.

Left: Eggs in worker cells.

Development times

Queens, drones and workers have different developmental times and it's important that you learn these as they have a direct bearing on the understanding of various manipulations, particularly swarm control. They can be seen in the table below.

Days	Queen	Worker	Drone
0	Egg laid	Egg laid	Egg laid
3	Hatches	Hatches	Hatches
5		Diet changed	Diet changed
8	Cell sealed		
9		Cell sealed	
10			Cell sealed
11	5th moult (prepupa)		
13		5th moult (prepupa)	
14			5th moult (prepupa)
15	Final moult		
16	Emerges		
20	Mature	First moult	
21		Emerges	
23			Final moult
24			Emerges
25	*		
27	Mates		
29		Flies	
31	*		
35		Mature	
37			Mature
41	Too old to mate	Foraging begins	

For the first three days all larvae are fed royal jelly, which consists of secretions from the mandibular and hypopharyngeal glands. After this time the diets of the worker and drone are changed to brood food, which contains a higher percentage of hypopharyngeal gland secretion and a component derived from pollen.

At the fifth moult the larva changes to a prepupa and metamorphosis to the adult form begins.

The starred lines in the table indicate the limits of the timeframe during which the queen can take her mating flight.

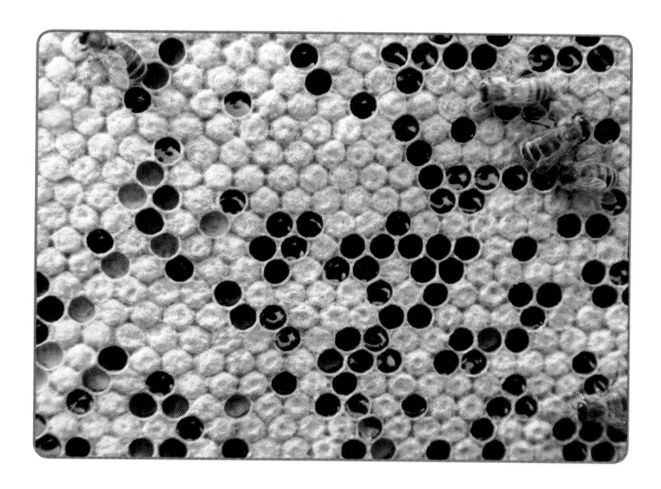

Above: Sealed worker brood.

Worker and drone cells

Workers, drones and queens develop in cells specific to them. Worker and drone eggs are laid in the hexagonal cells of the comb, the majority of which are 'worker' cells. As the name implies, it's in these that the queen lays worker eggs. When the larva is due to pupate, the cell is sealed with a flat or slightly domed beeswax cap.

Being larger, drones need bigger cells in which to develop. Drone cells are often found at the edges of the comb. They're sealed with a more domed cap which is obvious when compared with those of worker cells.

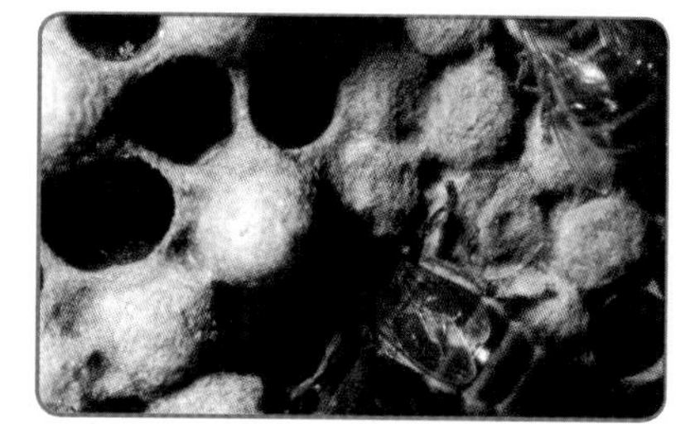

Right and below: Sealed drone cells.

Queen cells

Queen cells are completely different and hang down from the face or edge of the comb. In the spring, workers begin to construct queen cell cups, which look like acorn cups. This is normal and does not indicate a colony is preparing to swarm until you see the sides of the cup being elongated and the queen lays an egg in the cell. Once the queen cell is thus 'occupied', the sides are drawn out further as the larva develops. The cell is sealed and the larva pupates. In due course, the adult queen chews round the end and emerges into the colony.

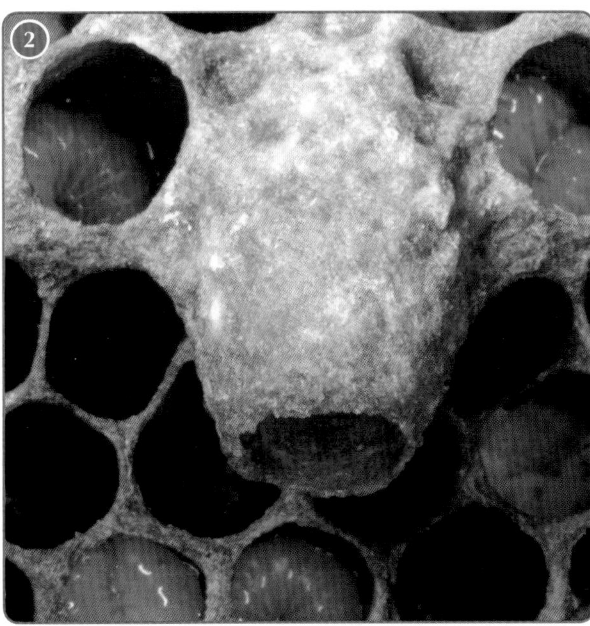

1 A queen cell cup.
2 An occupied queen cell being drawn out.
3 Sealed queen cells.
4 A virgin queen emerges.
5 A virgin queen on the comb.

Different types of bee

Bees come in different types or races, with subspecies within races, but all belong to the genus *Apis*, which is part of the order Hymenoptera. All Hymenoptera have two pairs of wings that can be linked together. Here we are mainly concerned with the bees that live in this country. As was mentioned earlier, our bee is the Western honey bee, *Apis mellifera*, which is represented by four major subspecies:

- ***Apis mellifera carnica*** – the carniolan bee, found from the Alps in Austria to Slovenia, Croatia and Serbia.
- ***Apis mellifera caucasica*** – the Caucasian bee, found in the mountainous area between the Black and Caspian seas, from southern Russia down to Azerbaijan.
- ***Apis mellifera ligustica*** – the Italian bee, originating in that country but since moved all around the world.
- ***Apis mellifera mellifera*** – the dark European honey bee of Northern Europe and Western Russia.

The last is the native honey bee of the British Isles, although it has been widely hybridised with imports of *Apis mellifera ligustica* and *Apis mellifera carnica*.

Each subspecies demonstrates different characteristics that it developed in order to survive in its original native area.

Apis mellifera carnica is greyish in colour and gentle to handle. It's resistant to brood diseases, although adults appear to be more susceptible to diseases such as Paralysis, Acarine (a mite which invades the breathing tubes) and Nosema (a gut infection in adults). It overwinters well and is thrifty with its stores. It builds up quickly in the spring but then swarms very readily. Because it comes from a wide geographical area, it has many variations or strains that can exhibit widely varying characteristics.

Below: *Apis mellifera carnica.*

Dr Kathy Keatley Garvey

Above: *Apis mellifera caucasica.*

Above: *Apis mellifera mellifera.*

Apis mellifera caucasica looks very much like the carniolan bee. It's also a gentle bee but doesn't have the same urge to swarm. Colonies are susceptible to Nosema and often die from this during the winter. Colonies do not build up quickly in spring. The Caucasian bee builds a lot of brace comb and collects large quantities of propolis, making hive manipulations more difficult.

Apis mellifera ligustica has yellow banding on the abdomen that varies from bright yellow to more of a leather colour. It's gentle to handle and can produce very large honey crops. It doesn't swarm excessively and builds good comb. However, when it's moved to areas where it experiences longer winters and lower temperatures than in its native area, it can have problems. Its tendency is to overwinter with very large colonies, which means a large amount of winter stores is required. Any additional food given to the colony is rapidly turned into brood rather than being stored. They can show low resistance to Acarine.

Below: *Apis mellifera ligustica.*

Apis mellifera mellifera is dark with long body hair. It overwinters well, is thrifty with its stores and collects large quantities of pollen. Colonies build up at a slower rate than other subspecies but individual bees are longer-lived and very hard-working, being able to fly further to forage. The bees produce very good comb with white cappings. Colonies can be strongly defensive but this is a characteristic that can be overcome with selective breeding.

A large proportion of the bees in the British Isles is hybridised because of the large volume of imports, largely of *Apis mellifera ligustica*-type bees from New Zealand and previously from the USA. However, recent research using DNA techniques has confirmed the presence here of pure *Apis mellifera mellifera* colonies.

From the above data it may seem like a good idea to purchase imported bees of a non-local subspecies, as they appear to offer the characteristics you want. However, before doing that think further into the future. Queens mate on the wing with drones that come to drone congregation areas from all the hives in the vicinity. These are very likely to be of different subspecies or hybrids between subspecies. While out-crossing can produce hybrid vigour, in bees it can lead to bad temper a few generations down the line if not immediately.

If you introduce a pure-race different subspecies into your area and your bees swarm, the virgin queen that the remaining colony produces will out-cross when she mates and your very docile, easy-to-handle colony could easily turn into a man-eater. However, if you obtain some local bees (checking that the colony is free of disease and easy to handle), when your virgin queen flies out to mate she's much more likely to mate with drones from colonies with similar genetics to herself, and the temper of her resulting colony is more likely to remain stable.

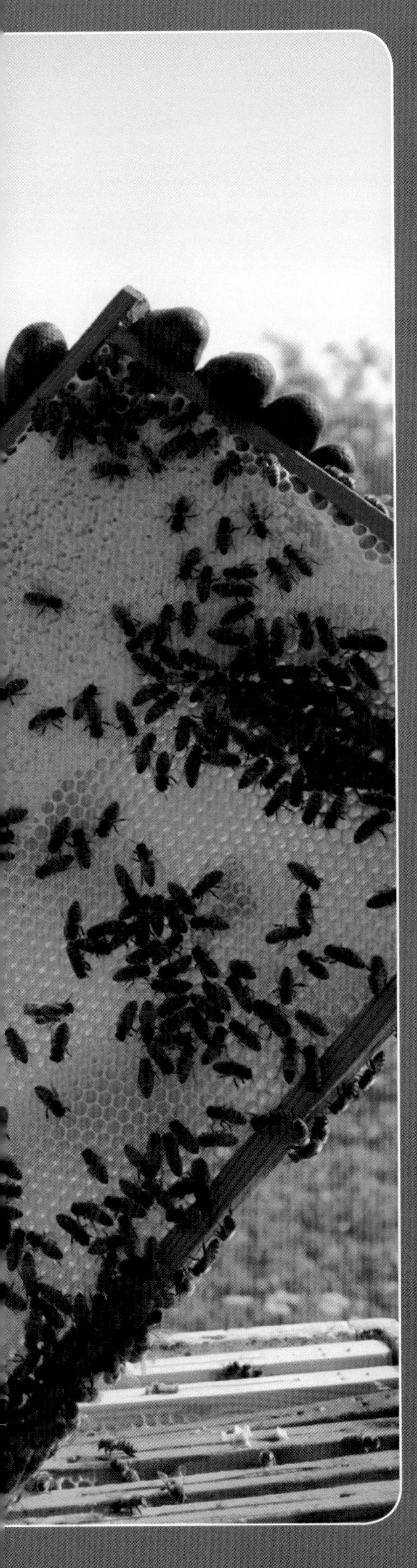

CHAPTER 2 BEE MANUAL

THE EQUIPMENT

Choosing a hive

Our UK bees naturally live in a cavity that protects them from the elements and, with a small entrance, enables them to defend their colony, and particularly the honey stores, from robbers such as bees from other colonies and wasps.

In some parts of Africa and Asia bees are still kept in log hives, but with this arrangement most management tasks are impractical or impossible. In Europe man took a different approach, and started out by keeping his colonies in straw or wicker skeps.

Above: Log hives in Africa.

The skep

Most skeps are made from a rope of twisted straw that is bound together to form a basket shape with a domed or flat top. There's either an entrance at the bottom or sometimes it's cut into the side or incorporated into a solid floor. A wicker skep is the same shape as a straw skep but woven from willow. This obviously leaves lots of gaps, so it then had to be 'cloomed', that is plastered over with a mixture of mud and cow dung. This made it both bee-tight and largely waterproof.

Above: A straw skep.

Some beekeepers still keep bees in skeps although this makes inspection for brood diseases and treatment for varroa more difficult. Nowadays skeps are mainly used for catching swarms. A well-made skep is light and extremely robust and a good one should support a man standing on it.

You'll find a lot of information about skeps at www.martinatnewton.com/page2.htm

Movable frame hives

In 1845–50, Reverend Dr Johann Dzierzon discovered that bees build their combs a fixed distance apart, allowing them just enough space to pass between them. Then in 1851 Reverend Lorenzo Lorraine Langstroth discovered that they

Below: Natural comb is built a fixed distance apart.

leave a space of 6–8mm (¼–5⁄16in) – the 'bee space' – between the edges of their combs and the cavity wall. This discovery changed the way that we keep bees because it allowed the development of what we know today as movable frame beekeeping.

Bees in an empty cavity start building their combs from the top, making it impossible to examine them without destroying the nest. By introducing frames, Langstroth made it possible for a single comb to be removed, inspected and replaced. All modern hives, of whatever design, are based around the concept of the bee space.

The bee space

Bees leave two bee spaces between midribs of adjacent combs in the brood nest area, and one between those in the honey storage area. Frames are therefore spaced accordingly and a horizontal bee space is also left between the top bars in one box and the bottom bars in the one above. If the gap is larger than the bee space, bees will fill it with what is known as brace comb. Smaller gaps will be filled with propolis. It's therefore very important to make sure your frames and hive boxes fit together correctly as this will make your life as a beekeeper very much easier.

Top or bottom bee-space hive designs

Hive designs are described as either 'top bee-space' or 'bottom bee-space'. In top bee-space hives, the bee space between boxes is above the top bars in the lower box, *ie* the bottom bars are in line with the bottom edge of the box. Bottom bee-space hive designs have top bars level with the top of the box and the bee space between the bottom bars and the base of the box.

Above: Brace comb.

Below: Propolis on the inner cover.

Above: Top bee-space top bars, bottom bee-space top bars.

Below: Bottom bee-space bottom bars.

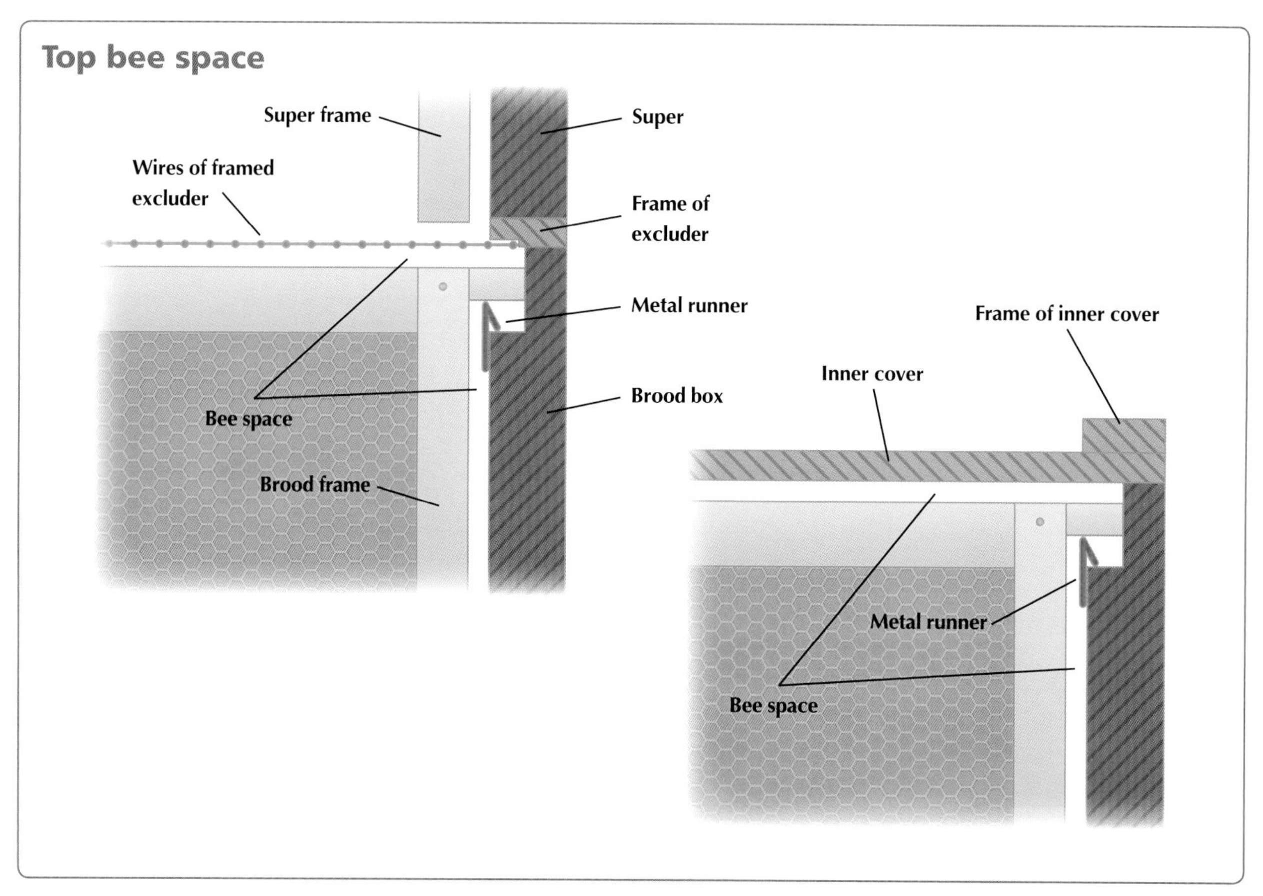
Top bee space
Super frame
Super
Wires of framed excluder
Frame of excluder
Metal runner
Frame of inner cover
Inner cover
Brood box
Bee space
Brood frame
Metal runner
Bee space

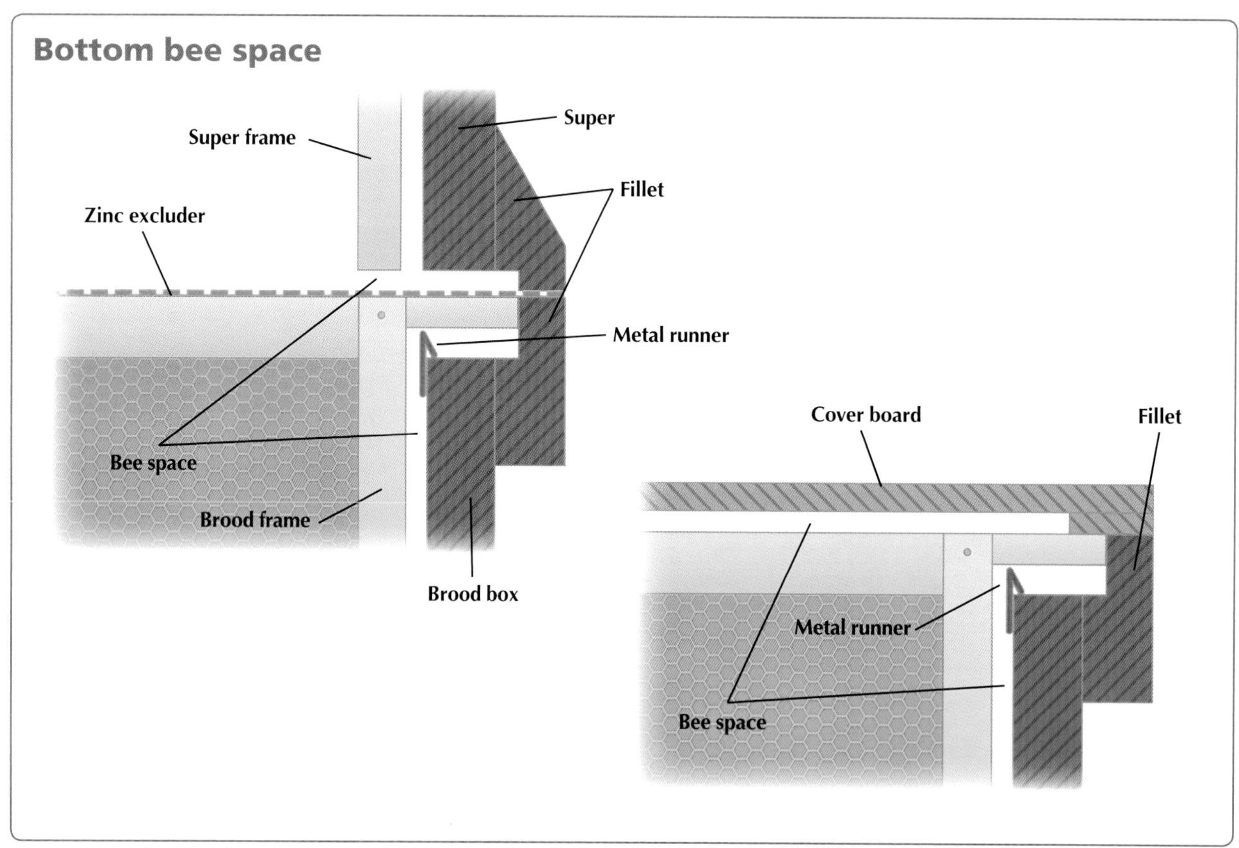
Bottom bee space
Super frame
Super
Fillet
Zinc excluder
Metal runner
Cover board
Fillet
Bee space
Brood frame
Brood box
Metal runner
Bee space

Do not mix top bee-space boxes with bottom bee-space boxes. If you put a bottom bee-space box on top of one with a top bee-space, you immediately create a two bee-space gap that the bees will fill with brace comb. The other way round, the bottom bars of the frames in the top box will rest directly on the top bars of those in the bottom bee-space box and the bees will glue them together with propolis. Both scenarios are bad news for you when you try to part the boxes to inspect the colony.

There are two important things to remember about a man-made hive. Firstly, the *internal* dimensions must be correct. Secondly, the hive must be 'bee-tight' – in other words, the only way in must be through the entrance. There must be no other gaps elsewhere large enough for bees from other colonies or wasps to gain access. If robbers can get inside, they can remove a large part or even all of the colony's honey stores, leaving it to starve.

Above: Robber bees gaining illegal access.

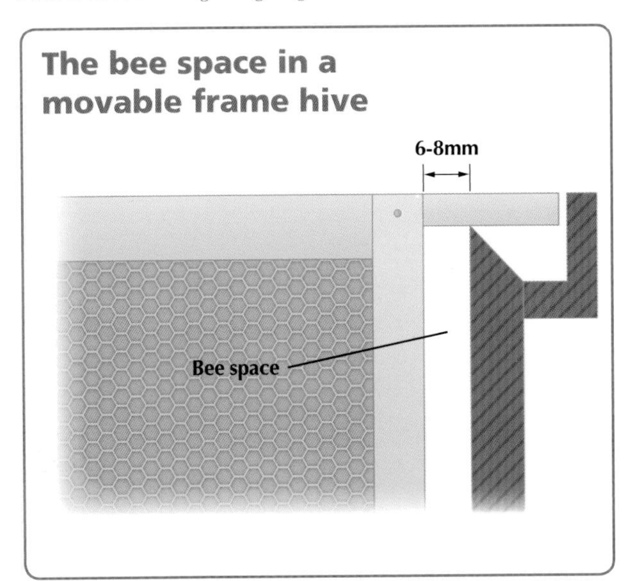

The basic hive

Modern hives essentially differ only in the volume of the brood boxes and supers, and whether they have top or bottom bee spaces. There is much debate as to whether top bee-space or bottom bee-space hives are better. I must confess I don't think it matters very much. As long as there's a single bee space between the boxes, the arrangement is the same as far as the bees are concerned.

All hives consist of the same basic parts. Working from the bottom upwards, these are:

- The floor or bottom board.
- The brood box.
- The queen excluder.
- The super(s).
- The inner cover or crownboard.
- The roof.

The floor

The floor is placed on a hive stand, which gives a comfortable working height and also reduces dampness. Until a few years ago floors were made of wood and were solid. However, with the advent of the varroa mite there has been a move to change to floors incorporating an open mesh panel and a removable tray which slides underneath, to enable monitoring of the level of natural mite mortality. Dead mites and those knocked off the bees fall through the mesh on to the tray. The gap is such that they cannot climb back into the hive, thus reducing the mite population. More about using mite mortality to determine when to treat your bees can be found in Chapter 6.

Below: A solid floor.

Above: An open-mesh floor.

The floor has fillets on three sides, giving an entrance on the fourth that's formed when the brood box is placed on top. Most commercially available floors are deep, *ie* the fillets are 22mm (⅞in) deep. However, I recommend that you consider using a shallow floor that's roughly a bee space deep (6–10mm, or ¼–⅜in). This has several advantages: bees are less inclined to build brace comb below the bottom bars on to the floor, and the entrance gap is too shallow for mice to move into the hive in autumn when they're looking for somewhere warm. A deep floor will need to have a mouseguard to keep them out. This is a strip of metal that's pinned across the entrance, with 10mm (⅜in) holes drilled into it so that the bees can still fly in and out.

Below: A reversible floor, deep above, shallow below.

Below: A mouseguard.

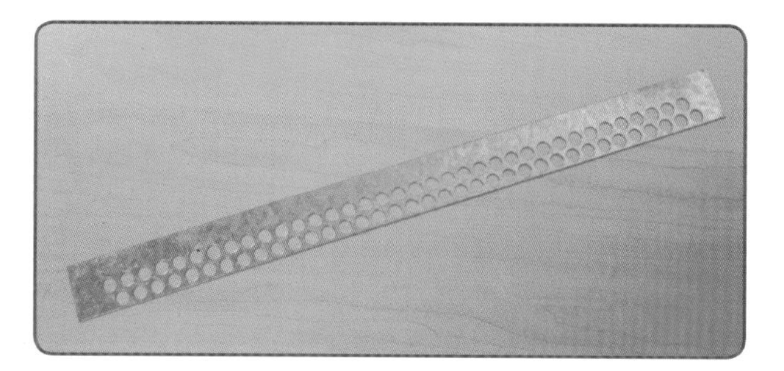

Above: The mouseguard in position.

An entrance block can be used to reduce the entrance in a deep floor. This fits the gap and has a depression, usually in the centre, measuring approximately 10mm (⅜in) high and 75–100mm (3–4in) wide. This gives an entrance that a strong colony should be able to defend easily. If the weather is very hot you'll want to remove the entrance block, so it should not fit too tightly.

Below: An entrance block.

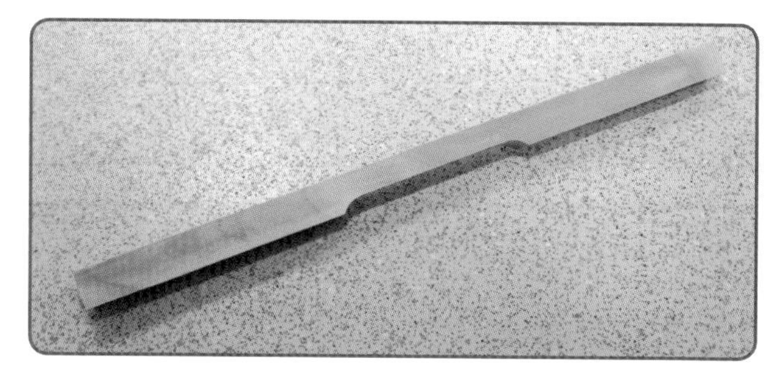

Below: The entrance block in place.

The brood box

The brood box stands on the floor. This is where the queen lays her eggs. Pollen and honey to feed the developing larvae are stored here. The brood box is a deep box containing frames supported on a pair of runners, fixed on opposite sides. Runners consist of a narrow strip of metal or plastic bent over along their length. This lip then curves over the thickness of the hive wall. The frames sit on the runners, which are positioned so that the top bar is flush with the top of the hive walls (bottom bee-space) or the bottom bars are flush with the bottom of the hive walls (top bee-space).

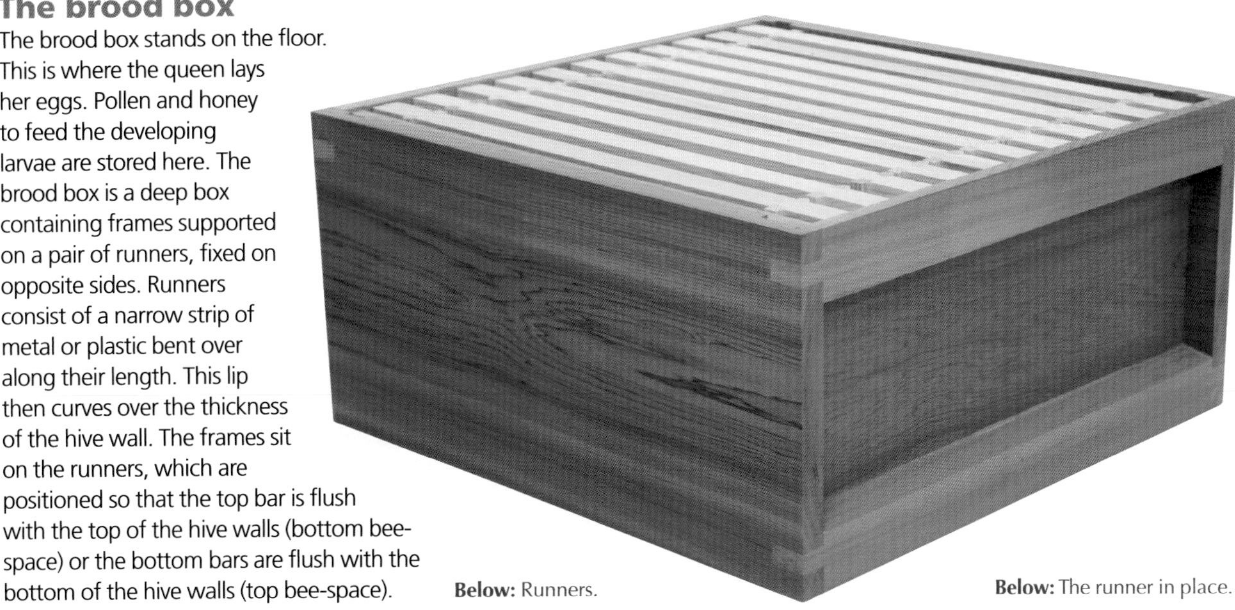

Below: Runners.

Below: The runner in place.

The queen excluder

Next comes the queen excluder. This is a framed grid or a piece of slotted metal or plastic the same dimensions as the cross-section of the hive. Worker bees can pass through the gaps but, because they're larger, drones and the queen cannot.

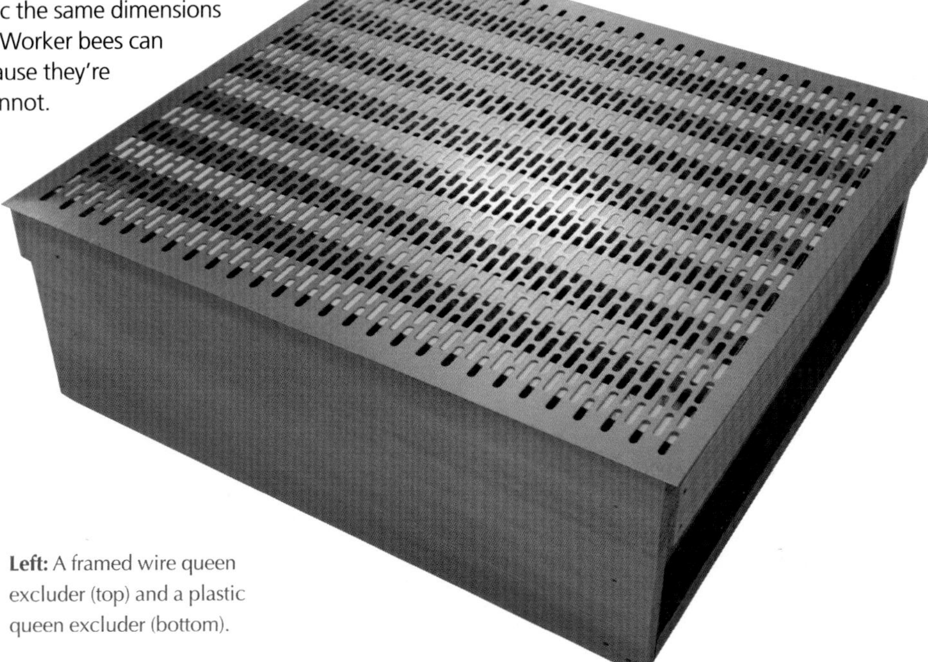

Left: A framed wire queen excluder (top) and a plastic queen excluder (bottom).

The supers

During the active season, foragers bring nectar back to the hive and store it as honey. The colony's natural inclination is to store honey above the brood nest. In a natural cavity this is achieved by extending the combs downwards, but the bees cannot do this in a hive. Instead, additional boxes, called supers, are placed on top of the hive, extending the nest upwards. Supers have the same cross-sectional dimensions as the brood box. They're generally shallower than the brood box and take shallow frames to fit their dimensions.

As a general rule, you need three or four supers for each of your colonies. Nectar is roughly 40% sugar and 60% water whereas honey has a water content of around 18–20%. Thus nectar needs more storage space than honey does. When you put a super on the hive, the bees will use this to unload nectar they've collected so that they can go out for more. The nectar will then be converted to honey, and in due course the super will be filled.

The inner cover or crownboard

The inner cover goes above the supers, or the brood box when there are no supers on the hive. It's a flat board with the same cross-sectional dimensions as the brood box and supers, generally has fillets round all four edges on both sides of the board, and usually has one or two holes cut in it that can be used when feeding the colony. It also doubles up as what's called a 'clearer board' by inserting Porter bee escapes (see page 112) in the holes. These are one-way devices that allow bees to leave the supers but prevent them returning.

Left: An inner cover with holes to take Porter bee escapes, converting it to a clearer board.

The roof

Unsurprisingly the roof goes on top of the hive. It has ventilation holes in the top centre of each side that allow air to flow through the hive, helping to keep it dry and to prevent mould growth. The holes are protected internally with mesh, to prevent wasps or robber bees gaining access. The roof is covered with a thin metal sheet or other suitable watertight material.

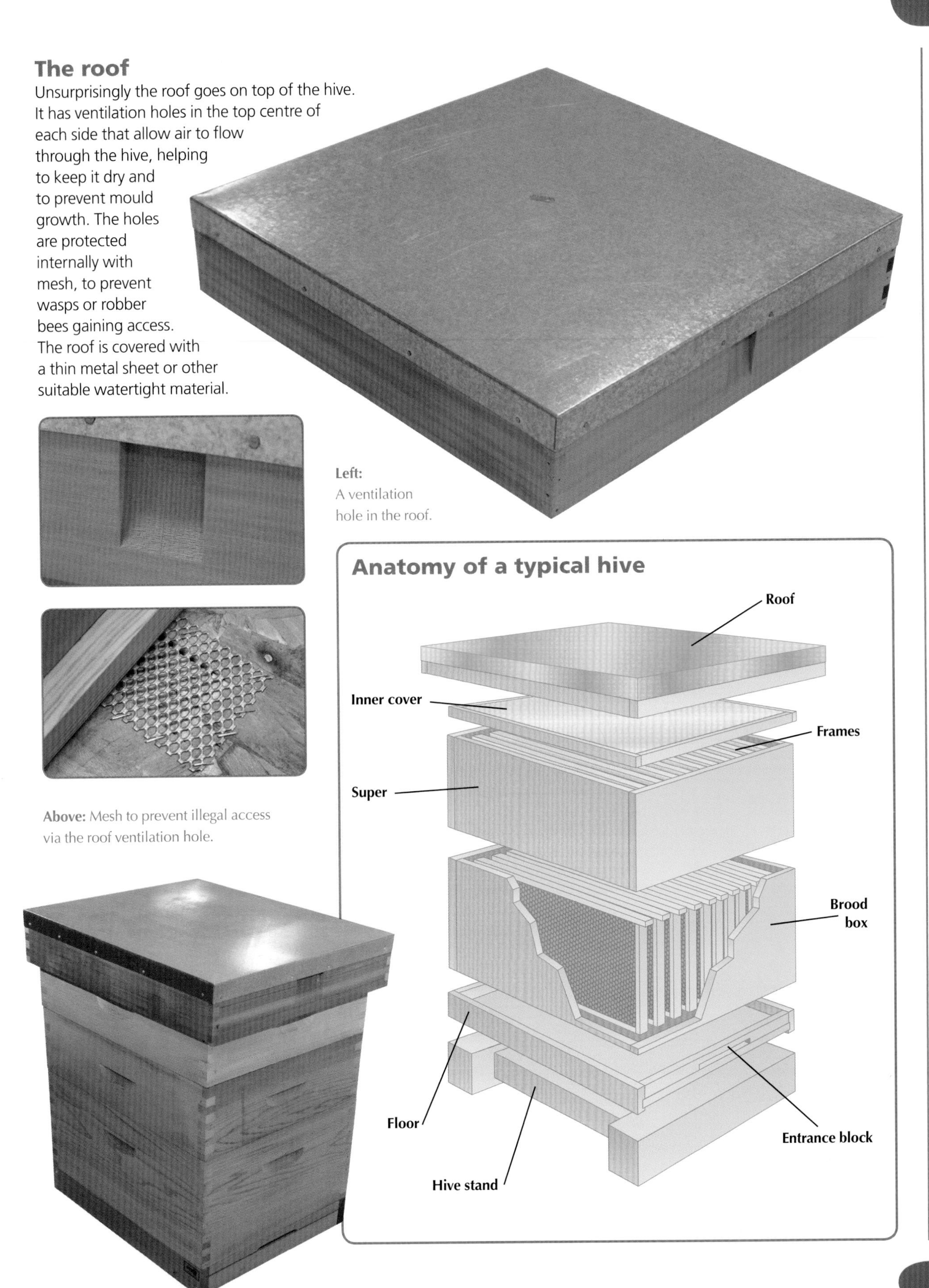

Left: A ventilation hole in the roof.

Above: Mesh to prevent illegal access via the roof ventilation hole.

Single- and double-walled hives

Most beehives are single-walled, *ie* they're made of single pieces of wood, nailed at the corners. They come in a number of different sizes. The Langstroth hive is the most popular single-walled hive in the world. However, most beekeepers in the UK use a smaller hive known as the Modified National or simply the 'National'. Other single-walled hives include the Smith, the Commercial and the Modified Dadant.

Above: A Langstroth hive

Below: A Modified National hive.

National Bee Supplies

Above: A Commercial hive.

National Bee Supplies

Above: A Dadant hive.

In colder areas you might consider using a double-walled hive. These have inner boxes made from less substantial timber, with secondary boxes known as lifts stacked round them and supporting the roof. In the UK the best-known double-walled hive is the WBC, designed in 1890 by William Broughton Carr. Most non-beekeepers recognise this as a 'typical' beehive with its pagoda-like structure and apex roof. The colony gets some extra protection from the elements from the lifts and, in severe weather, insulation can be placed between the lifts and the inner boxes.

Above: A WBC hive.

However, there are disadvantages to the WBC hive, as its double-walled construction makes colony inspections more time-consuming, as you'll have to lift off the roof and most of the lifts before you can start looking inside the internal boxes. With a single-walled hive you only have to remove the roof. In addition, if you knock the inner boxes when you're removing or replacing the lifts you could disturb the bees and transporting your hive any distance won't be easy. It's difficult to strap a WBC hive together firmly and safely so that it won't come apart during a journey and bees won't be able to escape. The additional wood required also makes such hives expensive.

Below: Inner boxes of the WBC hive.

Other hive types

The majority of hives used in the UK are made from wood, either Western Red Cedar or pine. Treated properly these will last many years. They can be painted with wood preservative (making sure it's one that doesn't kill insects) or paint. If they do get damaged during use, or a woodpecker makes holes in them, they can be mended easily by anyone with basic carpentry skills.

Hives made from polyurethane or polystyrene are now available, mostly only in the Langstroth size although a National hive made from polystyrene has recently been offered for sale. You may want to try these.

Other hive designs include the top-bar hive, the Long Deep Hive, the Beehaus and the Warré Hive. These are all workable designs but I would recommend that you don't try them until you've gained some experience of working with bees and have learnt the first principles of how they operate, as these hives require different management strategies and much of what you read in basic beekeeping books won't apply to them.

Below: Examining a colony in a WBC hive.

Choosing a hive

Remember that a colony of bees is happy to make its home in any suitable cavity, which can be a hollow tree, a chimney or even something like a dustbin! Movable frame hives are primarily for the benefit of the beekeeper, to make a range of manipulations possible.

Your choice of hive is entirely up to you. If you only plan to keep a couple of hives down the end of the garden the WBC may well suit your purposes. However, if you're thinking of a larger operation or if you're planning to move colonies for pollination, then the single-walled hive is much more suitable. Most single-walled hives have depressions cut in the outside face of each side that are used to lift the box. The depressions are, of necessity, not very deep and this means you're lifting the box using only your fingertips, unless you can hold it underneath. This can be difficult when you've a box full of bees and/or honey. The Modified National hive has a slightly different construction from other single-walled hives, with additional fillets top and bottom on two opposite sides of the box. These make it much easier to lift.

Above: A finger hold.

Above: The fillet on the Modified National hive.

Before choosing your hive, consider the local climate and the type of bee you want to keep. If you live in a warmer part of the country, with more prolific bees that build up to a large colony during the active season, you should think about using one of the larger-sized brood boxes. An alternative is to operate a 'double-brood' or 'brood-and-a-half' system, where you've two brood boxes or a brood box and a super, one on top of the other, with the queen having access to both. In a colder area or if your bees are less prolific (but no less productive), the National or WBC brood box may give your colony sufficient room.

Below: A 'double brood' hive.

Above: A colony on 'brood-and-a-half'.

Check with members of your local beekeeping association to find out which is the most popular hive in your area. If you follow suit supplies will probably be more readily available, and there's also likely to be a buoyant second-hand market.

Don't mix hive types

When you decide on a hive design, stick to it. All the frames will fit all the boxes, all the boxes will marry up with bee-tight joints, and they'll be compatible with the floor, inner cover and roof. You can be very sure that if you have a mixture of designs you'll never be able to find a part that fits properly when you need it urgently. It's worth noting, however, that the Modified National and the WBC use the same-sized brood and super frames, so these two types could be run together. The Smith hive also uses the same size frames but with short lugs. Frames from a Smith hive can be transferred easily to a National or WBC hive, but if you want to move them in the other direction you'll have to shorten the lugs before they'll fit. As I said earlier, do not mix top bee-space parts with bottom bee-space ones.

Top bee-space hives are the Langstroth, Dadant and Smith. Bottom bee-space hives are the National, WBC and Commercial.

Warm way and cold way

The National hive is square so the brood box can be placed on the floor with the frames running parallel to the entrance (warm way) or at right angles to it (cold way). The warm way has a number of advantages, the main one being that when you're inspecting the hive you're standing behind it, keeping out of the way of the activity at the entrance. Bees are often reluctant to draw out comb just inside the entrance and with frames arranged the warm way this means that only one comb is affected. With frames the cold way, the front corner of a number of frames may not be drawn out.

Above: Frames the 'warm' way.

Left: Frames the 'cold' way.

Left: Examining a colony with the frames parallel to the body.

The brood nest is roughly spherical, although some bees like a tall, elongated shape. It's enclosed in a layer of cells containing pollen, with the rest of the comb being used to store honey. The combs slice through the sphere. With frames the warm way, the brood pattern is symmetrical, with the brood nest area in the middle. If the frames are cold-way, the brood nest is generally found towards the entrance side of the frames, with the pollen above and honey stored towards the back of the hive.

Above: The brood nest in a 'warm way' colony.

Left: The brood nest in a 'cold way' colony.

The winter cluster forms around the brood nest and bees maintain the colony's temperature by consuming their honey stores. As these are eaten, the cluster naturally moves upwards to more honey stores. It's not natural for it to move sideways, especially if doing so means crossing a gap between frames. A colony on frames the cold way may therefore starve in the winter even if it has honey stores on either side of the brood nest. This is known as isolation starvation. The general rule is that bees should always have stores above the cluster when you've fed them for the winter, regardless of which way you have the frames.

Below: A colony which has died from isolation starvation.

Frames and foundation

As explained earlier, the discovery of the bee space made it possible to introduce frames that could be removed and replaced in the hive. Fitting these frames with a sheet of wax ('foundation' – see below) encourages the bees to build their comb within the frame – for the beekeeper's convenience!

Frames

Frames are made to fit the brood and super boxes of the different hive designs. The parts are usually sold separately in packs of ten. This is inconvenient if you're using the National hive, which takes eleven frames, but I'm sure you'll soon find a use for your spare ones. If you're part of a beekeeping group you could pool your order to make it more cost-effective.

A modern frame consists of a top bar, two side bars and two bottom bars, which slot into two grooves at the base of the side bars. The top bar has a removable wedge that's used to secure the sheet of foundation in place. Side bars come in three main types, all of which have a groove on the inside face into which is slid the sheet of foundation.

Frame spacing

The first type of frame has straight narrow side bars and is designated DN1 (deep) or SN1 (shallow). Spacers ensure that these sit the correct distance apart. The commonest form of spacer is the 'plastic end', which slips over the frame lug. The ends butt up against each other in the hive, spacing the combs. 'Narrow' plastic ends are used in the brood box where combs need to be two bee spaces apart. In the supers, where only one bee space is required, frames can be spaced more widely if required, using 'wide' plastic ends, so that the bees increase the depth of the honey cells.

Left: A narrow plastic spacer.

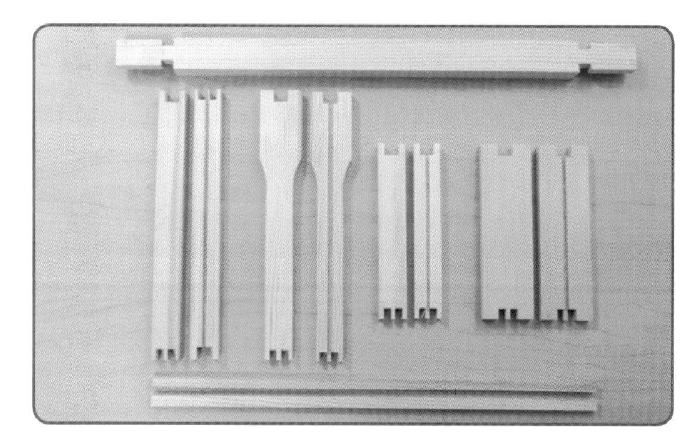

Above: Frame parts: top bar, different side bar styles and bottom bars.
Below: A made-up frame.

Above: Narrow plastic spacers in place.
Below: Wide and narrow plastic spacers.

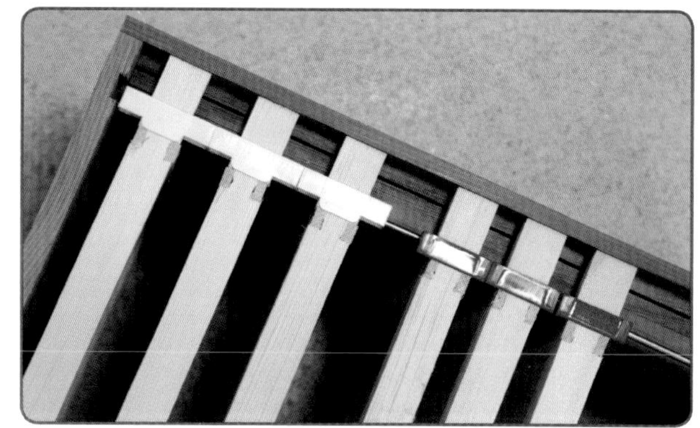

Self-spacing frames

Self-spacing or Hoffmann side bars make life much easier, certainly in the brood box. From experience I would recommend these rather than plastic ends, but all will work. The Hoffmann side bar is the same width as the DN1/SN1 side bar at the bottom, but just over halfway up it widens so that frames are held the correct distance apart. One edge is flat and the other is chamfered. This gives a narrow point of contact between adjacent frames, which minimises propolisation. With the groove for the foundation on the inner face of each side bar, adjacent frames will match up properly.

Left: The Hoffmann self-spacing side bar.

Below: Interacting Hoffmann frames.

Some beekeepers use Manley side bars in the super. These are the same width along their length spacing the super frames correctly. This stops the combs swinging if the box is moved, particularly if colonies or super boxes are transported by vehicle for pollination or honey collection purposes.

Castellations

Frames can also be spaced using castellations. These are flat metal strips with square notches cut into them to accommodate the frame lugs. Castellations come with nine, ten or eleven slots, depending on the hive type and whether you want narrow or wide frame spacing. They're nailed to the inside of the box, ensuring that the tops of the frames are correctly positioned for top or bottom bee-space. It's important to get this correct or you'll have problems with brace comb or propolis.

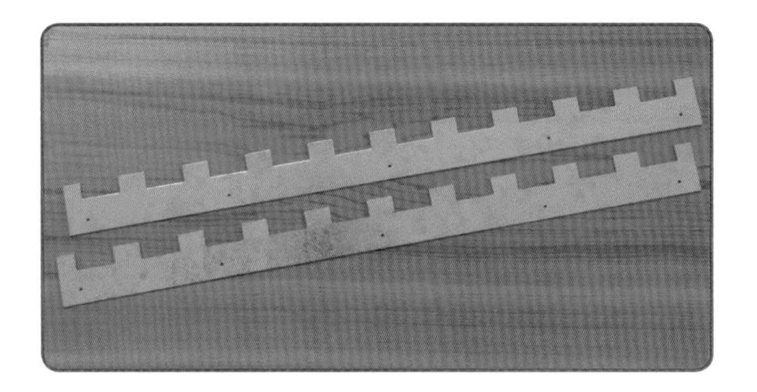

Above: Castellations.

Below: The castellation in place.

Although some beekeepers use castellations in the brood box, I would advise against this as it makes manipulations more awkward.

Using Manley or Hoffmann frames in the supers means you'll do better using a radial rather than a tangential extractor. As the width of these side bars holds the comb further away from the cage in a tangential extractor, there's a greater chance that the combs will be broken during extraction (see page 116).

Wide and narrow top and bottom bars

Top and bottom bars are both available as standard (narrow) or wide. Although slightly more expensive, I would recommend using the wide version of each as this cuts down the construction of brace comb between the frames, again making inspections and manipulations easier.

Above: Wide (upper) and narrow (lower) top bars.

Above: Wide (left) and narrow (right) bottom bars.

Making up frames

It's very important to make up frames accurately. They must be square and flat or the bee space will be compromised and your life will be more difficult than it need be. If you have a number of frames to make up, it's probably worthwhile making a jig that will ensure the top bar, side bars and bottom bars go together at right angles. If you don't have a jig, then use a set square to check before finally nailing the frame together.

Before you start making up a frame check that you have all the parts and tools to hand. You'll need a hammer, a craft knife, a jig or set square and a pair of pliers (for removing wonky nails). You'll also need some frame nails, which can be purchased from equipment suppliers.

First, I recommend a dry run. Take the five parts of your frame (top bar, two side bars and two bottom bars) and fit them together. Make sure the grooves in the side bars are

Above: Equipment for making up frames.

facing inwards, as these will later hold the foundation in place. If you're using Hoffmann frames this will also ensure that the vertical edges butt up correctly when you put them in the hive.

The top bar has a piece attached to it with a thin sliver of wood. This is the wedge. Remove it and shave off the excess sliver with your craft knife. Don't lose the wedge, as you'll need it later to fix the foundation in place.

Above: The top bar showing the wedge.

Next, push the two side bars on to the top bar. Use the set square to make sure that they're at right angles to the top bar and then secure each to it with two nails, one being driven in from each side.

Below: Making sure the frame is square.

Take one of the bottom bars and push it into the grooves at the bottom of the side bars on the side opposite the missing wedge. Make sure each end is flush with the outside edge of the side bar, making the frame square at the bottom as well as the top. The bottom bars give the frame rigidity and restrain the sheet of foundation in a vertical plane. They don't take any weight and should not be nailed in the same way as the side bars. Nail each end from underneath so that the nail points up the length of the side bar rather than across it. This will make it easier to remove the bottom bars when you come to replace the foundation after a few years.

Left: Nailing the bottom bars.

You can make up frames at any time and store them in a brood or super box (making sure the wedges are kept safe). However, I recommend that you don't fit the foundation until shortly before you want to put the frames into the hive. Wax is plastic (malleable) and sheets of foundation can easily warp. Once this has happened you can't flatten them out again, and if you give them to your bees they'll build wavy comb. This is not a problem for them as they'll keep the faces of the combs the correct distance apart, but it makes manipulations much more difficult, especially if you wish to move the position of a frame within the box or transfer it to a different box where its profile won't fit that of the other combs.

Foundation

Empty frames could be put into a hive and the bees allowed to build their comb wherever they wanted. However, this will usually result in comb being built across the frames rather than separately within each frame, so a means had to be devised to persuade the bees to build comb where the beekeeper wanted it. The solution is a sheet of 'foundation' to go into each frame.

Foundation is a sheet of beeswax, cut to fit the appropriate frame size. It has a hexagonal cell pattern embossed on each side, offset in the same manner as the bees would produce cells naturally. Bees presented with this ready-made basic structure will use their beeswax to draw out the embossed pattern into cells, thus making combs within the frames.

Wired foundation

In the UK, foundation is generally sold as 'wired' in order to make it stronger in the brood box and, in supers, less likely to collapse in the extractor. In the commonest wiring pattern a single wire is embedded into the wax in a vertical zigzag pattern with large loops left at each turn at the top edge and smaller loops at the bottom. Another pattern has six or seven crimped wires embedded vertically into the sheet with the ends sticking out slightly at each side.

Above: Wired foundation in the frame.

Unwired foundation

Unwired foundation is also available. Some beekeepers prefer to put two or three lengths of wire across their frames rather than having it embedded in the foundation. They then insert a sheet of unwired foundation and warm the wires slightly to embed them into it. Thin unwired foundation is used if you want to produce 'cut comb', where sections of the entire comb are cut out. The wax and honey can then be spread directly on to bread or toast and everything is eaten.

Above: Cut comb honey.

The easiest way to obtain foundation is to purchase it from an equipment supplier. After you've been keeping bees for a while and have collected your wax from cell cappings or combs that you've melted down for some reason, you'll be able to part-exchange it for new foundation or even use it to make your own. However, until then your bees will have to build comb from scratch on the foundation you provide.

Storing foundation

Keep your foundation flat and horizontal until it's needed. Place it on a flat board slightly bigger than the sheets, with another board on top. It can help to put a small weight on top too. Keep it in a relatively warm (but not too warm) place. If it's kept too cold it will become brittle, and if it's too hot it will start to melt. If you can get permission, one of the best places for it is the airing cupboard, which has a warm, constant temperature.

Inserting foundation

Assuming that you're using wired foundation and slide the foundation carefully up the grooves in the side bars. You may need to trim the sheet slightly in order to make it fit snugly. If your frame is slightly out, the foundation will not slide up smoothly. Then bend over the protruding wires/large loops at the top so that they lie in the right angle of the top bar where you removed the wedge. This operation is very much easier if you can grip the top bar in a vice of some kind, with the side bars pointing upwards.

Above: Inserting foundation into the frame.

Replace the wedge, which will now cover the wire loops and the top part of the foundation. Secure the wedge with three frame nails driven through into the top bar. These can go through the wire loops, but that's not essential.

Above: Nailing the wedge in place.

Finally, fit the second bottom bar, making sure that the foundation falls in the gap, and secure it in place as before.

Above: A complete frame with wired foundation.

Plastic frames and foundation

Some beekeepers use plastic foundation, which is often supplied ready fixed to a plastic one-piece frame. The embossed plastic sheet is dipped to give it a thin covering of beeswax, which makes it more attractive to the bees. However, some bees don't readily take to plastic foundation and are reluctant to build out comb from it.

Above: A plastic frame and foundation.

Above: Comb built on plastic foundation.

Plastic frames and foundation can be sterilised and recoated if necessary. The comb in wooden frames can also be replaced, and it's good practice to renew foundation every two to three years to reduce the risk of disease.

Renewing foundation

To re-foundation a frame, first cut out the old comb, wrap it up to stop bees finding it, and dispose of it. It can be burnt and any residue composted. Remove the wedge and clean out any wax. Scrape the wax out of the grooves (you can buy a handy tool which makes this easier), remove one of the bottom bars and scrape both of them clean. You should sterilise frames before reusing them. Use a blowtorch to lightly scorch the wooden surfaces or scrub them with a strong solution of washing soda. If you have access to a 'Baby Burco'-type boiler, sets of frames can be boiled in a washing soda solution to clean and sterilise them.

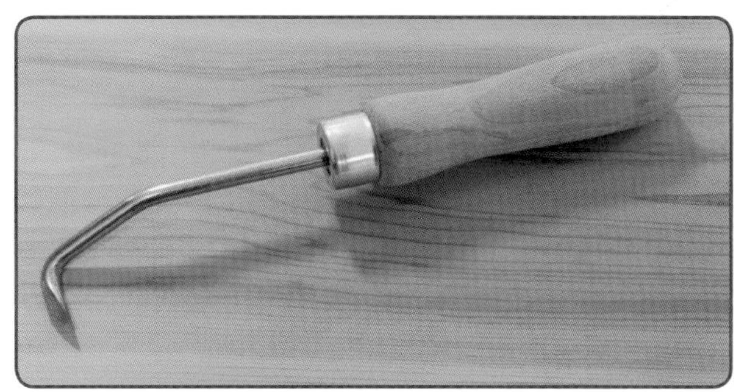

Above: A frame cleaner.

Left: Cleaning the groove in the side bar.

Below: Equipment for sterilising frames.

Once your frames are clean, insert a clean sheet of foundation as described previously.

Purchasing drawn comb

Be very careful when you purchase bees already on comb. Combs can be a source of disease and you should check them very carefully before you purchase a nucleus or full colony (see page 62). I strongly recommend that you do not buy spare frames containing drawn comb. If you must buy second-hand frames and comb, then replace the comb with new foundation after sterilising the wooden parts.

Assembling hives and frames

You can purchase a beehive already assembled and ready to go with frames and foundation already put together. This is certainly one of the quickest ways to get started, but it's also one of the most expensive.

Most beekeepers therefore purchase hives and frames 'in the flat', in other words in a pack containing all the necessary pieces sawn accurately to size and shape for your chosen hive design. This hive pack will also include runners or castellations as appropriate, sufficient nails and instructions. A recently available DVD takes you through the construction of a flat-pack hive (see www.bee-craft.com/shop).

Making your own

If you're skilled at woodwork, you may consider producing and assembling all the hive parts yourself. Plans for the WBC, Langstroth and other hive parts can be found on the Internet at www.beesource.com/build-it-yourself/, or the British Beekeepers' Association sells plans for the WBC, Modified National, Smith and Langstroth hives (see www.britishbee.org.uk). However, as it's essential that the internal dimensions are correct and that the boxes are square, I would recommend that you don't try this unless you're very handy with a hammer and saw. It's also likely to be a more expensive and certainly a more time-consuming way of making your hive. As well as producing the wooden parts, you'll need some 50mm (2in) 2.7mm (⅛in) galvanised nails for the main construction. You can strengthen the joints by applying waterproof glue before nailing them together. To finish the hive, you'll need a pair of runners for the brood box and runners or castellations for the supers.

Traditionally, Western Red Cedar was used to make beehives as it doesn't warp and doesn't need a preservative. However, it's expensive, and most hives today are made instead from pine, which is durable, particularly if you treat it with a preservative – but it's very important to ensure that the preservative doesn't contain insecticides, or you'll harm your bees. You can buy special hive paint, or linseed oil is also suitable.

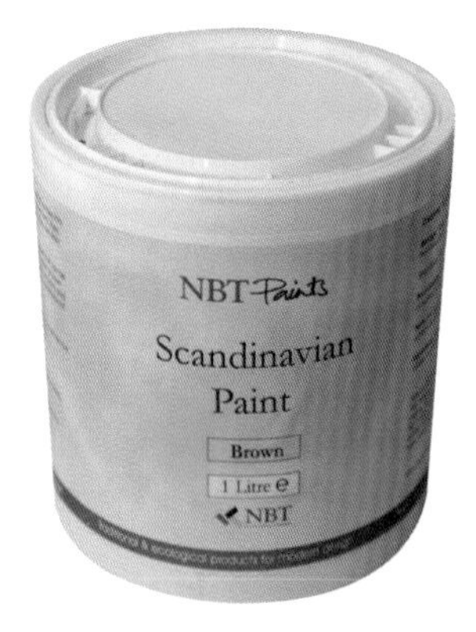

Above: Hive paint.

Your hive will take a lot of knocks during its life so it needs to be made from timber at least 19mm (¾in) and preferably 22mm (⅞in) thick. After that it doesn't really matter how thick it is because it's the internal dimensions that are critical. These must be accurate to maintain the bee space.

Assembly

Whether making your own or assembling a flat-pack, you must make sure that the corners of the boxes are at 90° and the whole thing is square. The side walls must be flush at the corners and all the boxes need to fit together, in any combination, without any gaps. Particularly with a single-walled hive, you need to be able to interchange all the boxes. A jig or long sash cramps are useful to hold the box square before fixing with nails. Pre-drilling the nail positions at the corners will minimise the chance of the wood splitting. If you're making up a National box, I would suggest making up the two walls that carry the fillets first and then joining them to the other two sides. Be careful you position the bottom fillet correctly so that there's a bee space under the bottom of the side wall.

Above: Making sure the box is square.

Left: Getting the bottom bee-space correct.

Positioning runners or castellations

With the box assembled, you need to attach the runners or castellations. For the National hive these go on the sides with the fillets. On any other hive they fit on the rebates on two opposite walls.

To get the runners or castellations in the right position, put a frame in place at one end of the box and adjust the runner's position for top or bottom bee space before nailing it to the side wall. Don't bang the nail right in at this stage as you may need to adjust it when you move the frame to the other end of the box to position the runner there. You can put a frame in one of the slots in the castellation and do the same. When the runner/castellation is in place and the frames correctly positioned, fix it firmly to the side wall. Repeat for the opposite wall.

Above: An open mesh floor with a removable insert.

Above: Top bee-space.

Above: The grid of an open-mesh floor.

You must also make sure that the floor, inner cover and roof are all square and flat. Any warping in these parts will lead to gaps and your hive will no longer be bee-tight. Open-mesh floors are available from equipment suppliers, but if you want to make your own you'll need black epoxy-coated mesh with eight mesh per 25mm or inch. The mesh prevents the house bees from removing the mites but it also stops them clearing out other debris that accumulates. This can be a breeding ground for wax moths so it's important to clear the trays out regularly, even if you're not monitoring mite levels.

Cover ventilation holes in the roof with mesh to prevent illegal access. Protect the roof with metal or another waterproof material and make sure that the cover folds down over the sides and is nailed in place.

Right: A complete Modified National hive.

Choosing an apiary site

Bees choose their nest site using criteria that have been developed over many millennia of evolution. These include the volume of the cavity and the size and orientation of the entrance. What they don't do is to choose a home site because it's near other bees. Bees also tend to choose cavities that are 3m (10ft) or more from the ground. From their point of view, an apiary is entirely unnatural.

Shelter from the wind

That being said, an apiary must be placed so that it causes the least harm and trouble for the bees. One of the most important criteria is shelter from the wind. Bees can only fly under control when they're moving faster than the air around them and can only land easily when flying 'into the wind'. Logic says that hives should not be sited facing the prevailing wind. A thick hedge will slow down the wind speed along the ground for a distance of 40 times the height of the hedge.

Bees blown down by cold winds will never recover enough to get back to the hive and are lost to the colony. A bee whose body temperature falls to 9–10°C (48–50°F) cannot move sufficiently to raise that temperature and will die.

Below: Hives protected from the weather by a hedge.

Shelter from the sun

During the autumn, winter and spring, colonies can benefit if their hives are exposed to the sun. A small amount of warmth on a hive can make the difference between bees remaining in a winter cluster or flying out and being able to defecate. However, those in the same position during the summer can overheat. A practical answer is an apiary site shaded by deciduous trees that lose their leaves in winter.

Even here in the UK, unshaded hives can get so warm inside that large colonies are more likely to start swarming preparations and even issue as a swarm earlier than normal, thus negating the value of regular colony inspections.

Below: Hives kept in a deciduous area.

Orientation

Given the option, bees will choose a nest with the entrance facing roughly south. If your apiary site is well sheltered from the wind, in practice this doesn't matter too much and you can orientate hives to suit your situation.

However, all of the above suggestions are merely recommendations, and must be adapted to your own particular circumstances. Apply your common sense!

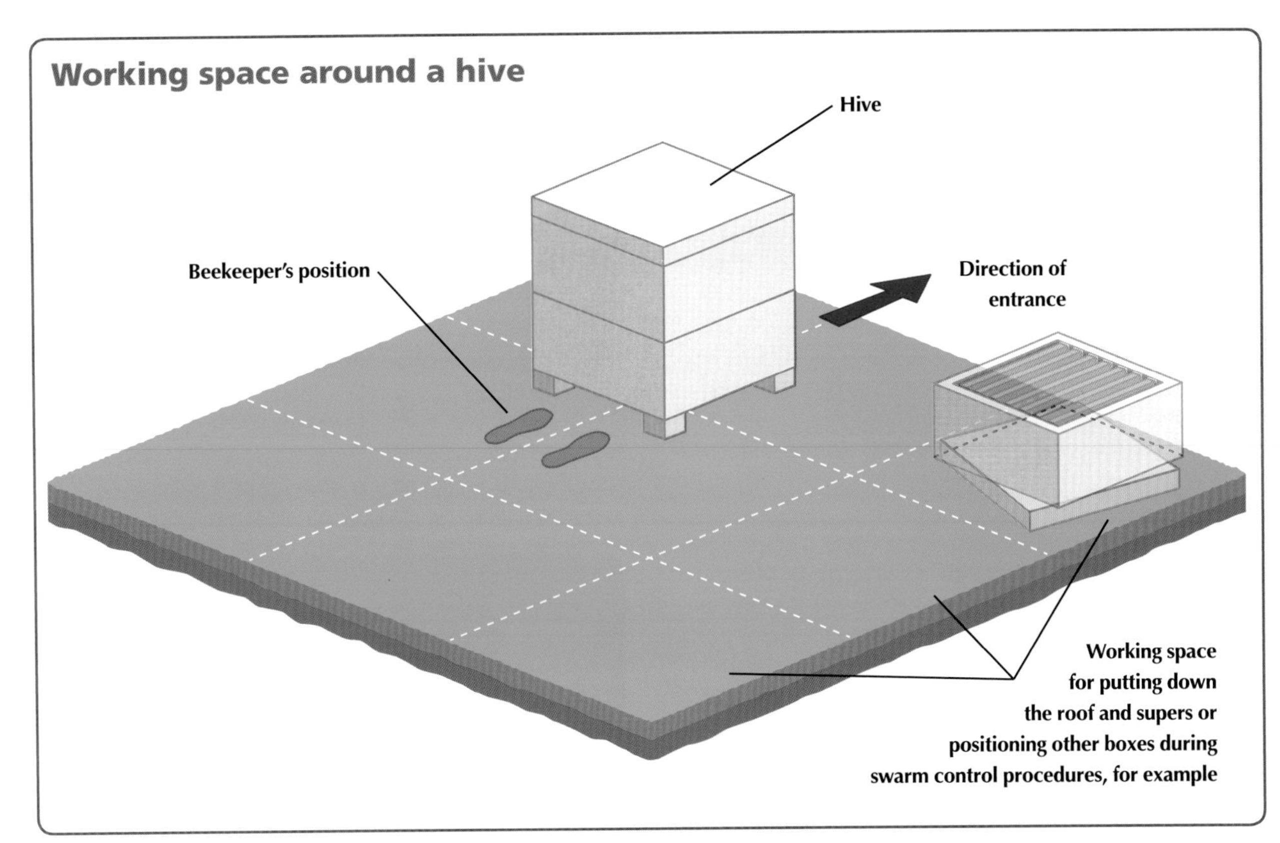

Hive arrangement

Within your apiary site, the arrangement of the hives can be important. Bees are confused by regular repetitive patterns and consequently may enter, or drift into, the wrong hive. Drifting is one of the main ways of spreading diseases.

When you inspect a colony, you'll remove the roof and place it upside down to one side of the hive. Then you'll remove any supers and stand them on it. You therefore need sufficient space around the hive to do this and continue your inspection. As a general rule you need to allow nine times the hive footprint area per hive, although this is not an absolute measure, as keeping two hives side by side reduces the overall area required. Make sure that your apiary site is large enough to accommodate the maximum number of colonies you plan to keep and for any boxes that may be needed as temporary accommodation for bees during swarm control manipulations. Like other things, bees tend to follow Murphy's law and several will probably decide to swarm at the same time.

In an apiary of two hives they can be placed side by side, facing in the same direction some 0.6–0.9m (2–3ft) apart. You should have room to stand so that the frames in the hive are parallel to your body. This means behind the hive for warm-way frames and to one side for cold-way frames, so that your hands can move naturally to the

Above and below: Manipulating warm-way (top) and cold-way (bottom) hives.

Above: Colonies in a large circle, facing inwards.

Above: A block-and-rail hive stand.

frame lugs. If you can only stand behind a cold-way hive you'll find it extremely awkward to have to reach across the hive to the front to grasp the lug, and lifting the frame out will also be more difficult than necessary.

Larger numbers of hives and stands can be arranged in many ways to break up any semblance of uniformity. The more hives you have, the more important it is that you do this to prevent drifting. One pattern is to place the hive stands in a large circle with all the entrances facing outwards or inwards. I think this reduces drifting to an acceptable minimum. Another favourite arrangement is to site eight hives in a block with two facing out on each side of the square. Place the stands far enough apart so that you can enter the square and manipulate the hives easily. Remember that you also need space to put down the roof and supers from the colony being examined.

Hive stand design

Hive stands need to be strong. A hive with full supers can weigh over 100kg, so the stand should be strong and stable enough for your fattest friend to stand on!

There are lots of different designs. One of the simplest is to support two strong square-section rails on concrete blocks. Alternatively you can fix a square-section frame to four sturdy legs. Placing the legs on flat tiles or slabs will stop them rotting. Some stands are meant to take a single hive while others will accommodate two or more standing side by side. I find that putting two hives together works well. This requires a stand about 1.5m long, as you need to be able to place the hives some distance apart, near each end. You can then face your 'twin' hive stands in different directions or you can plant small shrubs around them to serve as landmarks and make each site look different.

Your hive stand must also be level, both from front to back and side to side. Bees in a level hive will build vertical comb in the vertical frames whereas if the hive is badly tilted the comb will be vertical but it will not coincide with the

Above: A purpose-built hive stand.

frames. If your apiary site is sheltered rain should not generally blow into the entrance, except maybe in severe weather. Some beekeepers tilt their hives slightly towards the entrance to ensure that any water that's blown in can run out again. If you want to do this, place a 5mm block under the back edge of the hive.

Below: Protecting the leg of the hive stand.

Below: Raising the back of the hive.

Apiary sites

The fact that bees can be kept in a wide variety of places makes beekeeping an activity that can be enjoyed universally. One of the most obvious locations is your back garden, which you can modify to make life easy for the bees and comfortable for your neighbours. Everyone has the right to 'enjoy their property', and while there's generally no restriction on where you can place bees on land you own, your bees and your beekeeping must not interfere with the rights of your neighbours in their homes. This means that with the best will in the world, a tiny garden is not the best environment for beekeeping. Allotments can be good, but you need to check whether you need permission to move bees in there. Some allotments welcome bees while others either have heavy restrictions or ban them altogether.

Bees can also be kept on flat roofs and, indeed, this is becoming more and more popular in London and big cities, where several hotels now have their own hives on the roof and offer their own honey in the restaurant. A rooftop site means the bees are flying well above the heads of neighbours. However, the drawback here is the wind. You'll need to shelter the hives, perhaps with pot plants, a bamboo screen or something similar. Watch the bees and decide whether you need to provide further screening to make it easy for them to fly in and out. One of the things you need to watch particularly is that you have sufficient space. It's no good filling all the space with occupied hives. You'll need to move around and be able to get to the hives to inspect the colonies without falling over the edge.

Out-apiaries

Many beekeepers keep their colonies away from their homes in what are known as 'out-apiaries'. These can be in a farmer's field, an orchard or another spare piece of land belonging to someone else. If you can use the corner of a field, you'll only have to erect one fence to make the apiary stock-proof. You must keep livestock away from beehives.

Since you'll have to travel to get to an out-apiary you'll obviously need vehicular access to the site. In the long run, moving everything in and out with a wheelbarrow or a hive barrow is simply not practical. Consider whether your vehicle is suitable for moving hives. I once chose a particular vehicle because it could accommodate six hives in the back!

It helps to know your potential landlord. Building a relationship is one of the best ways of making sure you retain the site and, if it's a good one, such efforts will be worth their weight in gold. You can offer to pay the rent with honey or anything else that is mutually agreeable. Any site can be improved and, particularly if you're on good terms with the owners, you may get permission to grow or erect windbreaks, etc.

If you're going to your out-apiary alone, take your mobile phone with you and turn it on. Even with docile bees, you're likely to get the occasional sting and this could lead to an allergic reaction when least expected. In a very few cases this reaction can be fatal. If you react badly to a sting, with restricted breathing, a bad rash and swelling, you must get help as fast as you can.

Below: Protecting the apiary from livestock.

Water

Bees need water. Scout bees will search the area for a suitable source and then return and recruit other foragers. Although you and I know that the bees are merely drinking and have no evil intentions at all, non-beekeeping neighbours don't necessarily see the situation that way. What they see is a lot of bees round the edge of their pond or at a damp patch under a dripping tap, and they may be frightened. You can help to avoid this problem by providing your bees with water near to the hives but ideally a short distance away from them, as bees will naturally look for water further afield rather than very close to home.

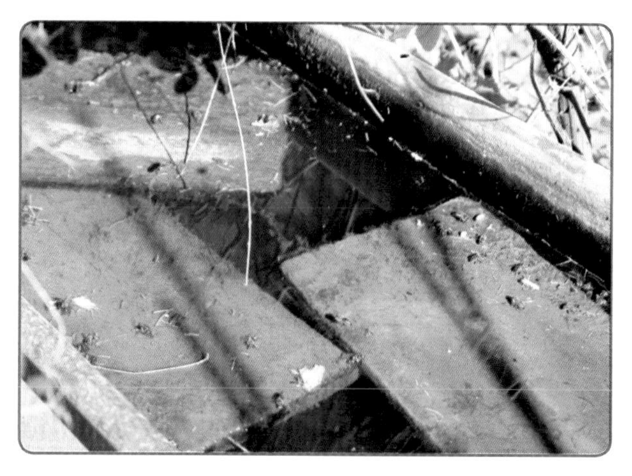

Above: A water fountain.

Left: Bees drinking from a leaking tap.

Below: Wooden floats so that the bees don't drown.

In a shallow container the water will warm up quickly on a sunny day. However, so that your bees don't drown in a deeper container you need to give them landing places from which they can drink. These can be stones that break the water surface or floating pieces of wood. Another approach is to make holes in the bottom of a bucket and fill it with peat or soil. Stand the bucket in a larger watertight container and fill this with water. The water will be sucked up to the surface of the soil and bees can land safely and drink.

The most important thing to remember is that once the bees start drinking from one of your water sources, you must make sure that it doesn't dry out. Check it regularly and top it up as required.

To encourage your bees to use your water source rather than one in your neighbour's garden, you should 'bait' the water when you establish a new source. Fill your source initially with a 1:1 sugar:water mixture. Trickle a little of this over the top bars of your colonies to encourage the bees to hunt for its source. Once they've found it, top it up with water. As you keep doing this, the solution will become more and more dilute until it's simply water. As long as you keep it topped up the bees will continue to drink there.

Neighbours

Sometimes neighbours are not as fond of your bees as you are and you need to be aware of this, and remember that they have a right to enjoy their garden just as much as you do yours. You need to take this into consideration, particularly if you want to keep your bees at home. If you're on good terms with your neighbours, talk to them before you get your bees and tell them of your exciting plans to become an apiarist. Explain that you'll need to open the colonies every seven days, especially if your work pattern is such that this will inevitably be at weekends.

Problems may arise in the spring and at swarming time. Bees are confined to the hive during winter and their bodily waste builds up in the rectum. On the first fine spring day they'll fly from the hive to defecate, but unfortunately the yellow-brown faeces can land anywhere. Fine spring days are also good for doing the washing, and you'll not be popular if your neighbour's clean white sheets get covered in bee excreta!

A swarm that leaves your hive may cluster anywhere, and this could be in your neighbour's garden. Worse than that, it might decide to take up residence in next door's chimney or cavity wall. The best solution here is to take steps to prevent and control swarming in your colonies. However, you might not always be successful so be prepared to go and collect a swarm as a high priority.

Any hedge above normal head-height will divert bees above passing humans and reduce potential contact or conflict. A high hedge will also hide your hives from sight, and if your neighbours cannot see them they'll often be unaware that the bees are even there. This may well solve problems that are imagined rather than real.

Left: Facing a colony into a hedge to make sure bees fly up and away from neighbours.

Problems with neighbours can also be reduced if you ensure that you always keep gentle bees. As a minimum, you should be able to walk around the apiary without a veil and not be bothered by the bees. If you don't have bees buzzing round your head in this situation, it's unlikely they'll be bothering the people next door. You need to keep bees that don't follow you after you've closed the hive after an inspection and don't defend their hive at a distance.

If there's a hint of trouble I'd strongly advise you to move the bees elsewhere as soon as possible. Members of your local association may be able to help, even if only with temporary accommodation for your colonies. Trouble with bees and neighbours is like a broken candle. Whatever you do, it cannot be completely mended.

Below: Consider your neighbours when deciding where to keep your bees.

Protective clothing

There are certain pieces of equipment that are essential for the beekeeper. The first is the veil. This is designed to protect the face from stings, and you're strongly advised never to go near an open hive without one.

The veil

Veils come in various designs. The simplest is a cylinder of black net drawn together at the top with elastic. Black enables you to see through it easily. The net goes over the crown of a brimmed hat and hangs down to just below the shoulders. Tapes at the bottom pass under the armpits to make the veil secure.

A more usual variation of the net veil is the 'hat and veil'. Here the net is sewn to the wide brim of a cotton hat, and a metal or stiff plastic ring is usually sewn into the net at about chin level to keep it away from the face. Once again tapes or elastic secure it under the armpits.

These types of veil should be worn with the bottom edge on top of the shoulders to make it bee-tight. If the net comes over the shoulders and the top of the arms, this creates a gap through which bees can enter.

Above: A 'Sherriff' veil.

Below: A simple hat and veil.

All these veils are normally worn with an ordinary cotton boiler suit that protects the rest of the body from stings, propolis and other dirt.

Just over 40 years ago a new type of veil was designed and made by Brian Sherriff, and today all veils of this type are known as 'Sherriff' veils. The concept was an all-in-one veil. The veil itself was a new design, being a hood rather than a cylinder. The back of the hood is material and the black mesh veil at the front is supported by two rigid semi-circular arches to keep it away from the face. The veil is attached to either a jacket or a full suit with a zip, enabling the hood to be thrown back from the face and removed completely when the suit part is washed. These veils now come in a variety of designs, materials and colours and you're advised to look for the one that suits you best.

The 'Sherriff'-type veil probably offers the best protection, but this can, in some ways, be too good. With full protection you may be unaware of the temperament of the colony you're inspecting and not take steps to rectify bad temper by re-queening.

Gloves

I recommended you wear gloves to protect your hands from stings, particularly when you're just starting. These can be as simple as a pair of rubber washing-up gloves. Leather ones with gauntlets to protect your forearms are also available. Some beekeepers like to wear a pair of thin disposable latex gloves, either alone or on top of another pair, to keep hands or gloves clean and free from the persistent smell of venom, which can provoke a reaction the next time a colony is inspected.

Boots

Some bees attack the ankle area (and hence are known as 'ankle biters'). Wearing rubber Wellington boots provides protection. The boots will also give you good grip, so that you don't slip while carrying a heavy box of frames, for example.

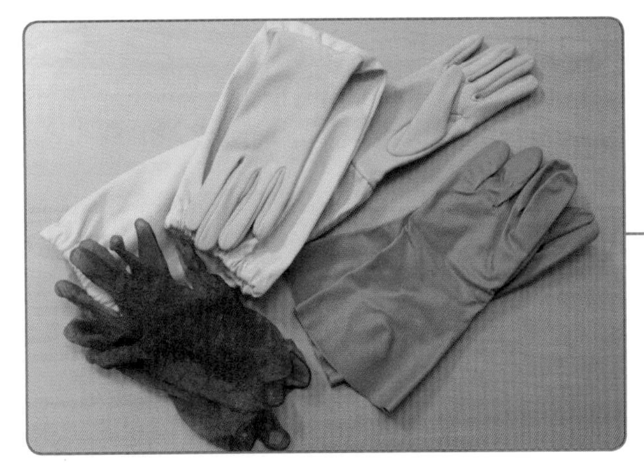

Above: Suitable gloves range from leather with gauntlets, through washing-up gloves to thin latex which may be worn over another pair.

Below: Rubber Wellington boots protect the ankles from stings and also give safe footing in the apiary.

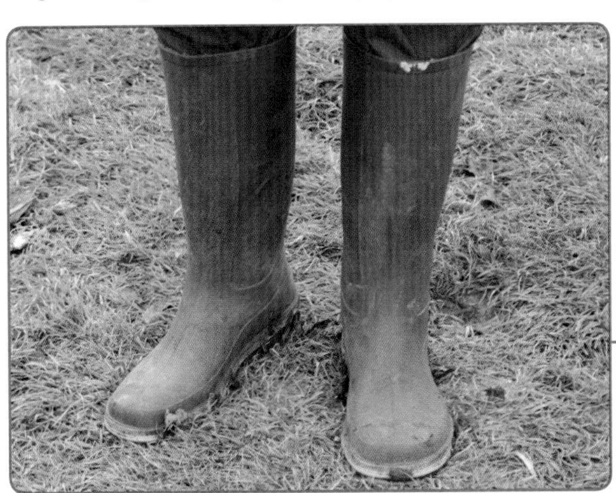

Right: When not in use, the hood of a 'Sherriff' veil can be thrown back.

Smokers, hive tools and record cards

As well as protective clothing, you need two other essential pieces of equipment: the smoker and the hive tool.

The smoker

The smoker is used to control the bees in a colony. In the wild, bees made their nests in hollow trees and other cavities. When there was a forest fire, the smoke indicated to the bees that they should abscond and find another, safer, cavity elsewhere. Before leaving, they would engorge on their honey stores to provide fuel for the flight and the wherewithal to produce beeswax with which to build combs in their new home.

The theory is that the smoker provokes the same instinctive reaction. When a bee engorges, it's less able to curve its abdomen round to use its sting. A gentle application of smoke can also be used to move bees off an area of the comb, making inspection easier.

The smoker consists of a firebox with a nozzle and a set of bellows. A grid at the bottom of the firebox allows the draught to be forced up through the fuel to expel the smoke through the nozzle. The spring-loaded bellows expel air through a hole and short tube at the bottom, opposite a hole in the firebox.

When choosing a smoker, make sure the bellows produce a strong flow of air from the nozzle. Also ensure that the spread of the bellows is not too big for your hand to operate it comfortably. Once you've determined this, my advice would be to buy the largest one available. Wide-barrelled smokers are easiest to light, they last for the duration of the inspection, and they can be put out to conserve unburned fuel by plugging the nozzle for a while. Partly burned fuel lights more easily the next time. A larger smoker is also easier to keep going than a smaller one, and the last thing you want is to find the smoker has gone out just when you need it.

The smoker's structure

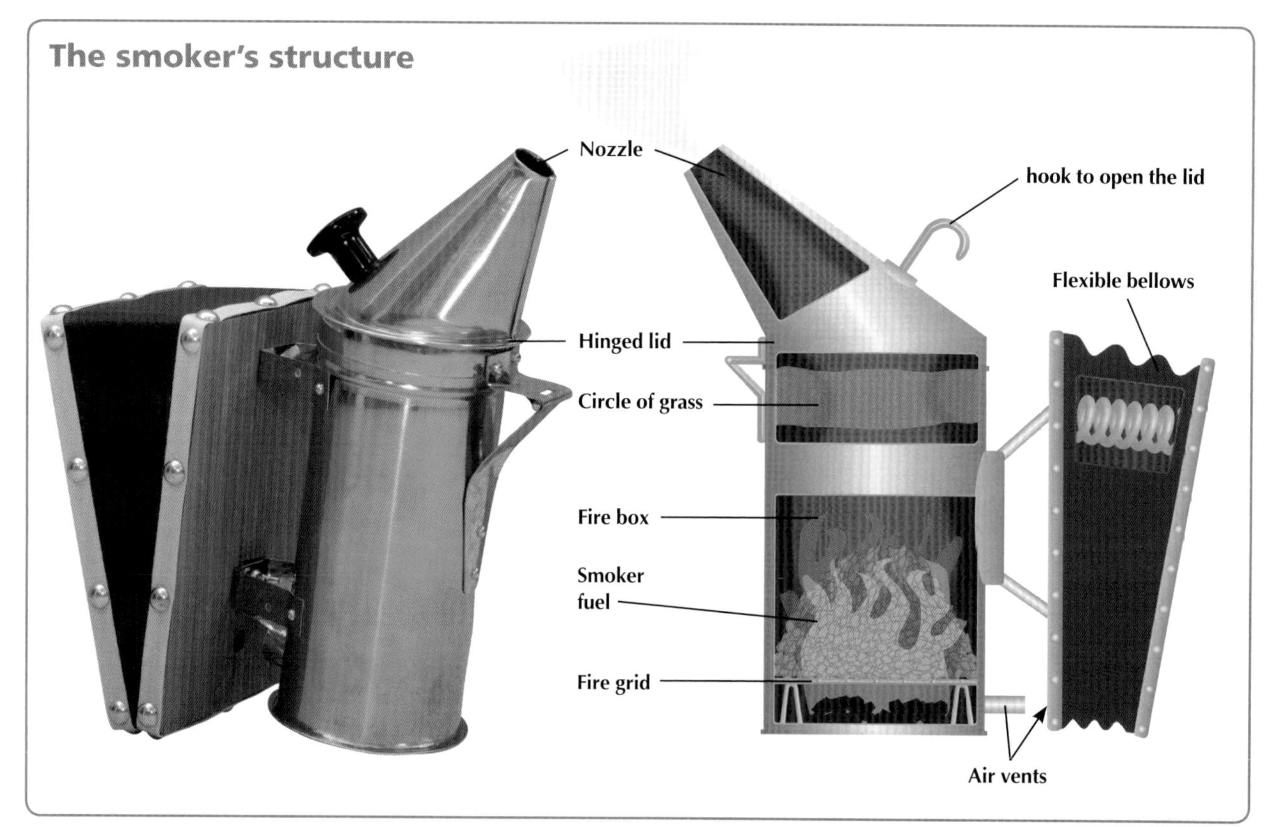

Lighting the smoker

Lighting the smoker and keeping it alight are things that are easy to say but often not so easy to do. The real test after lighting it is to put the smoker to one side and leave it. If it's properly 'on fire' it should still be capable of producing smoke 30 minutes later.

1 Tear off a piece of newsprint, about a quarter of a broadsheet page or half of a red-top sheet. Crumple loosely. Hold it over the top of the open smoker and light the bottom. Insert it into the barrel while puffing the bellows gently.

2 Continue to puff gently and start sprinkling in the shavings. Don't put in too many at first. You need to see flames at this stage.

3 Add more shavings and keep puffing. The aim is to build up a bed of red-hot embers. Add more shavings and keep puffing.

4 Fill the barrel to the top with shavings but don't press them down. You'll be producing lots of smoke.

5 Try leaving the smoker alone for ten minutes. Puff it to see if it still produces smoke.

Once you know you can do this every time, you're there! I have known a big smoker filled like this, then lit and left, still to be alight one hour later, with no puffing in between to keep it going.

If you're using a loose fuel such as planer shavings or Leylandii leaves, you can pick some long grass, twist it up and circle it inside the top of the firebox. This will help prevent pieces of smouldering fuel being expelled through the nozzle into the hive.

Smoker fuels

A number of different fuels work well in a smoker, but you're looking for one that produces a cool smoke. It also needs to be dry. Possible fuels are:

- Proprietary pellets.
- Machine planer shavings.
- Dead Leylandii leaves.
- Rotten wood (the type you can compress with your fingers).
- Well-rotted jute sacking.
- Corrugated cardboard.
- Dry pine needles.

If you choose sacking, make sure it hasn't been used to store something that may give off noxious fumes; and if you use cardboard, ensure that it hasn't been treated with a fire retardant. I suggest you start with shavings and try other fuels as you gain experience.

Don't make the first time you open your hive the first time you try to light your smoker. Practise beforehand. This means learn to do it right before you open your hive.

Extinguishing the smoker safely

To extinguish a smoker after your inspection, block the nozzle with either a solid bung (a flanged cork works well, or a piece of wood cut to size) or a twist of grass. It also helps to cover the hole at the bottom of the firebox, and grass works well here. Laying the smoker on its side also helps to put it out more quickly.

However, you need to be careful. The firebox will remain hot for a while, so don't put the smoker down where anything may melt or catch fire. Some beekeepers carry a tin box for the smoker in their vehicle, to prevent accidents.

Above: Putting out the smoker with grass plugs.

The hive tool

The second essential piece of equipment you'll need is the hive tool, used for a variety of tasks during inspections.

Hive tools come in two types: the 'standard' and the 'J'. Both have useful features and you should try both before deciding which you wish to buy. The standard hive tool has a flat edge at one end and is curved over at the other. Both ends are chamfered on one face. The 'J' tool also has a flat chamfered end, but the other end is curved in the same plane, giving its characteristic 'J' shape.

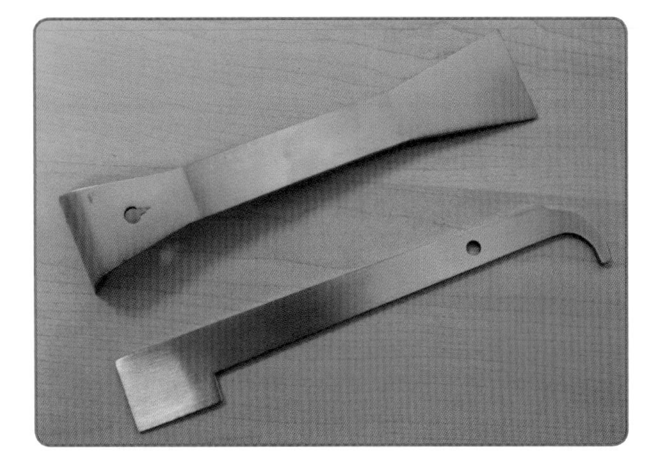

Above: The standard (upper) and 'J' (lower) hive tools.

The flat end of each type is used to separate boxes during inspections. Either end of the standard tool can be used to scrape wax and propolis from top bars, excluders and boxes. The hooked end of the 'J' hive tool is useful for raising the frame lugs so that you can get hold of them, especially the first frame you take out of the box. With the standard tool, it's probably easiest to use the curved end for this task but you'll have to decide which mode of operation suits you best.

Below: Using the hive tool to scrape the top bars.

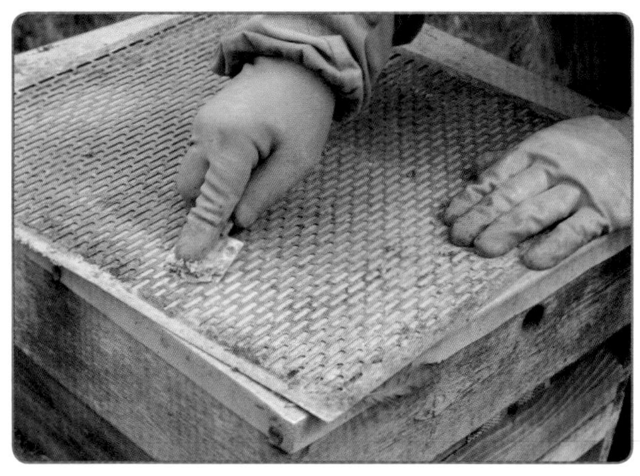

Above: Cleaning the queen excluder.

Above: Loosening the first frame.

Above: Using the 'J' hive tool to lift the frame lug.

Whichever hive tool you choose, it's essential that you learn to keep it in your hand all the time during colony inspections. You'll need to use it frequently and don't want to waste time picking it up or trying to find it. Attaching your hive tool to your belt with a bungee cord is useful in that if you do let go, you can retrieve it easily.

Keep your hive tool clean in order to reduce the risk of disease transmission, particularly between apiaries. Wax and propolis can be scraped off and the tool cleaned with a proprietary kitchen surface cleaner. Alternatively you can scrub it with a strong solution of washing soda.

The 'bee box'

You'll accumulate other gadgets and pieces of equipment as you go along so it's a good idea to keep everything together in a bee box – not, in this instance, a home for bees, but a practical storage container for all your bee-related tools and paraphernalia. This container should be sturdy and, if you're going to keep your smoker in it, needs to be made from a fire-resistant material. It will also need a handle to carry it, and it's useful to divide it into sections so that you can find things easily.

Below: The 'bee box'.

Colony records

A final essential piece of equipment is your colony records. Get into the habit of keeping these right from the start. At first, with only a few colonies, you'll probably remember what you saw and what you did to each one on every inspection. However, as time goes on, maybe with your number of colonies rising, remembering becomes more difficult.

Records are not just for checking what you did last time you looked at your bees, they're essential when you want to look back at a particular colony's performance or compare colonies at the end of the season in order to plan for the next one.

Colony identification

Some beekeepers number each set of hive boxes. However, I would recommend that you give each queen a number. This does not have to be physically attached to her, although you can get numbered discs which are attached to her thorax with a small dab of glue or paint.

Numbering the queen keeps a record of the most important bee in your colony. She's the one which gives her bees their characteristics, be it docility, a strong

Below: *Bee Craft* colony record cards include notes on the system and space in which to record medications given to your colonies.

The Bee Craft apiary guide to

Record keeping

How to keep records of your bees and their management

It is very important to keep records of your colonies and their management. This will enable you to apply management techniques such as swarm control and determine when you need to treat for disease, such as varroa.

The Bee Craft apiary record pack is suitable for an apiary of up to five colonies. It includes this instruction sheet, an example and facilities for recording veterinary medicines, five queen and colony record cards, five check lists.

Queen and colony record

- Allows the beekeeper to follow colony development during the season
- Records the work the beekeeper has done

Record details

- **Queen reference number** - each record card refers to a specific queen and the colony she heads and moves with the queen if she moves hives
- **Year queen raised** - in order to know the age of the queen
- **Queen mark colour** - you could use the conventional colours White/Yellow/Red/Green/Blue for years ending in 1/2/3/4/5 or 6/7/8/9/0
- **Clipped** - records whether the queen is clipped
- **Where mated** - records the apiary or location of mating
- **Current hive position** - can be used to sketch the position of the hive for each queen in pencil to allow for alterations if hives are moved
- **Date** - for the date of the inspection
- **Queen seen?** - Y for yes, N for no
- **Brood** - e for eggs, u for unsealed brood, s for sealed brood to track colony development
- **Food stores** - estimate the number of frames of stored honey or syrup available to the colony. If very low, consider feeding
- **Temper** - a colony which tries to sting, buzzes around your head and follows you home would score 0. One which makes no attempt to sting and continues working as if you were not there would score 5
- **Notes** - use this to record:
  queen cells - if present, how many and whether empty or occupied
  feeding - what food has been given
  mite monitoring - when a debris tray is inserted and removed and a record of daily mite drop
  medications - record of application and removal
  honey crop - honey or comb honey harvested

Contact Bee Craft on 01733 771221 or via the website: www.bee-craft.com

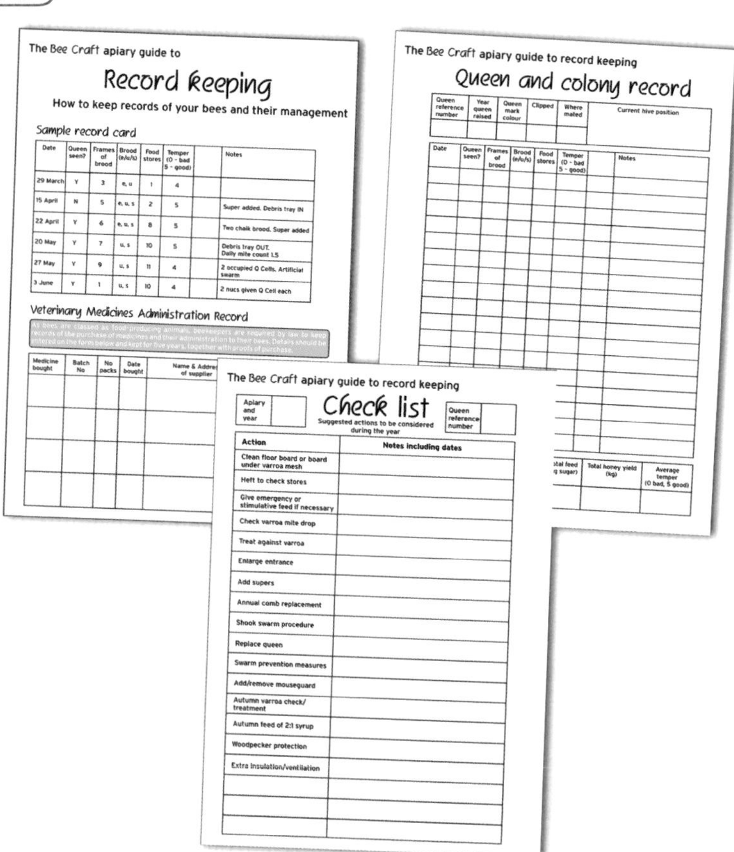

The Bee Craft apiary guide to

Record keeping

How to keep records of your bees and their management

Sample record card

Date	Queen seen?	Frames of brood	Brood (e/u/s)	Food stores	Temper (0 - bad 5 - good)		Notes
29 March	Y	3	e, u	1	4		
15 April	N	5	e, u, s	2	5		Super added. Debris tray IN
22 April	Y	6	e, u, s	8	5		Two chalk brood. Super added
20 May	Y	7	u, s	10	5		Debris tray OUT. Daily mite count 1.5
27 May	Y	9	u, s	11	4		2 occupied Q Cells. Artificial swarm
3 June	Y	1	u, s	10	4		2 nucs given Q Cell each

Veterinary Medicines Administration Record

Medicine bought	Batch No	No packs	Date bought	Name & Address of supplier

The Bee Craft apiary guide to record keeping

Queen and colony record

Queen reference number	Year queen raised	Queen mark colour	Clipped	Where mated	Current hive position

Date	Queen seen?	Frames of brood	Brood (e/u/s)	Food stores	Temper (0 - bad 5 - good)		Notes

The Bee Craft apiary guide to record keeping

Apiary and year

Check list

Suggested actions to be considered during the year

Queen reference number

Action	Notes including dates
Clean floor board or board under varroa mesh	
Heft to check stores	
Give emergency or stimulative feed if necessary	
Check varroa mite drop	
Treat against varroa	
Enlarge entrance	
Add supers	
Annual comb replacement	
Shook swarm procedure	
Replace queen	
Swarm prevention measures	
Add/remove mouseguard	
Autumn varroa check/treatment	
Autumn feed of 2:1 syrup	
Woodpecker protection	
Extra insulation/ventilation	

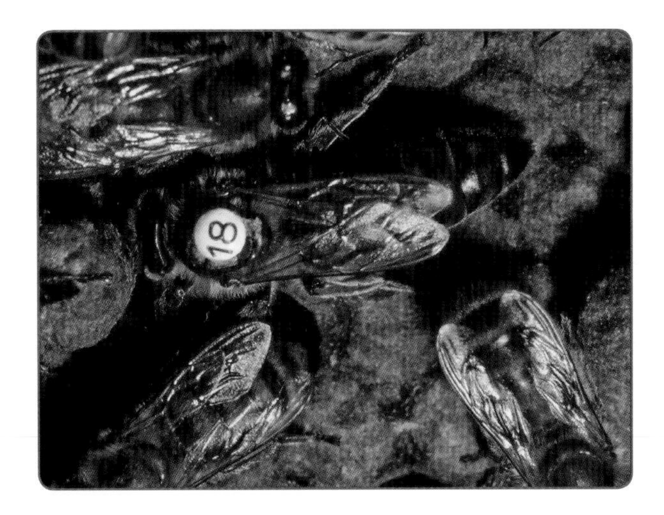

Above: A numbered queen.

Above: Tagging the hive with the queen identification number.

instinct to protect their nest, a propensity (or not) to swarm, or the ability to winter with thrifty use of stores. You'll want to assess characteristics such as these when you draw up your future management programme. If, for instance, your bees are particularly defensive (aggressive from our point of view), you may wish to change the colony character to a more docile one. The queen is the key to this operation.

It doesn't actually matter which particular box the queen is in. It's the queen herself that interests you. If, for example, she leads off a prime swarm and you catch it, this will end up in a different hive box but will have the same queen.

However, how you keep your records is up to you and the system should suit your situation. You could, for instance, simply keep a diary of each of your visits to the apiary. Here you can record what you saw and what you did in longhand, though in due course you'll probably find it easier and quicker to use one of the hive record systems that are available, or devise one of your own.

Bee Craft's Apiary Guides

Bee Craft, the Official Journal of the British Beekeepers' Association (www.bee-craft.com), produces a record card as part of its set of Apiary Guides. You use one card per colony. This has space on one side to record actions for each apiary visit. On the reverse is a list of general management tasks required during the year, which is both a useful aide-memoire and a way of recording when any tasks have been carried out.

The Apiary Guides would be a useful addition to your bee box. One gives pictures and descriptions of the stages of brood development (egg, larva and sealed brood) and of the queen, worker and drone. There are also sets illustrating the appearance of the major adult bee diseases, the varroa mite and methods of control, and exotic diseases not yet found in the UK. The last set describes the swarming process, one method of swarm control and how to collect and hive a swarm.

All of these, apart from the record cards, are laminated to help keep them clean and they fit in a handy binder to keep them together.

Below: The *Bee Craft* Apiary Guides.

CHAPTER 3 BEE MANUAL

COLONY MANAGEMENT

Acquiring bees

A lot of beginner beekeepers I have known have been desperate to acquire their first colony of bees. I know that I was, and I purchased a colony from the tutor of the beekeeping class I attended in the autumn. Looking back, that was probably not the wisest move as I put myself in the position of having a colony to look after when I had only attended half the lectures and had no practical experience whatsoever of inspecting a colony.

The course had two practical sessions scheduled the following April. That winter was particularly bad and, halfway through, I had to prevail on a beekeeper friend to come round to check that my colony had sufficient food to last it until foraging could start again. As it happened things were fine, but spring is certainly the time of year when colonies are most vulnerable and more likely to die of starvation. With hindsight, which is always so clear, I should have waited until I had finished the theory course, attended the practical sessions, and gone along to some of the apiary sessions organised by my local beekeeping association.

There are a number of different ways of acquiring your first colony. You can purchase your bees from a reliable supplier, either a recognised equipment supplier or someone from your local beekeeping association, or alternatively you can obtain a 'free' colony by collecting a swarm.

Buying bees

If you decide to buy, the first thing you need is a guarantee that your bees are free of major diseases. We will cover these in detail later, but there are four notifiable bee diseases: American foul brood (AFB), European foul brood (EFB), small hive beetle (SHB) and the *Tropilaelaps* mite. Small hive beetle and *Tropilaelaps* have not yet been found in the UK and we hope that this situation will continue, but if you know or suspect that your bees are infected by any of the notifiable diseases, you're duty bound *by law* to report this to the authorities. The relevant authority depends on where you live. In England or Wales you must report to the National Bee Unit, in Scotland to the local Scottish Government Rural Payments Inspections Directorate (SGRPID) Area Office, in Northern Ireland to the Department of Agriculture and Rural Development, and in the Republic of Ireland to Teagasc. Contact details can be found in the Appendix.

Virtually all colonies in the British Isles are infected by the varroa mite which, again, we will cover in detail later. However, you need the assurance that any bees you purchase have been treated for varroa at the appropriate time.

Standard for a nucleus colony

The British Standard for the make-up of nuclei (BS 1372) was withdrawn in 1984 but it still gives a good basis of what you should expect when buying bees. You should ask the seller if the nucleus complies with the criteria laid down in the British Beekeepers' Association (BBKA) information leaflet B14, *The BBKA Standard Nucleus Guidance Notes*. This can be downloaded from the BBKA website at www.britishbee.org.uk/files/nucleus_leaflet_2009.pdf. If there is any doubt, I would advise you to buy elsewhere.

The main criteria in the leaflet are:

- The nucleus should contain bees, brood, food and a queen of the current or previous season, reared in the UK.
- The queen should have produced all the brood in the nucleus.
- She should be marked with the 'standard' colour for the year in which she was raised (see later for the convention).

Left: A marked and clipped queen.

- The seller should clip the queen's wings at your request, if not already done so.
- The nucleus should have the number of frames stated.

Above: A five-frame nucleus.

- The frames should be securely pinned/nailed together and be sound, but they don't need to be new.
- Combs should be fully drawn, not foundation.

Above: Drawn comb on a sheet of foundation.

- There should be enough food to last the colony at least two weeks without any further food coming in.

Above: A food comb.

- A least half the total comb area should contain brood in all stages (eggs, larvae and sealed brood), with at least 30% being sealed and 15% being drone brood.
- There should be no active queen cells at any stage of development.
- The frames should be well covered with bees, with a good balance between young house bees and older flying/foraging bees.
- The bees should be good-tempered when handled by a competent beekeeper in suitable conditions.
- The brood and adult bees should be healthy with no signs of disease in any stage.

- Adult bees should show no obvious signs of damaged wings which could indicate deformed wing virus.

Above: Workers with deformed wing virus.

- A few cells showing chalk brood are acceptable as its presence can depend on the weather conditions.

Above: A few larvae dead from chalk brood is acceptable.

- The seller should tell you what treatments, including that for varroa, have been given and when they were administered. He/she has a legal requirement to keep such records and should be prepared to pass them on to you.
- Make sure all treatments given are legal, as once you purchase the bees you become responsible for any illegal substance that may be found in the honey or wax.

Your nucleus should be in a position to expand when you purchase it. This depends on there being sufficient bees present, a good amount of sealed brood that is nearly due to hatch, providing nurse and house bees for the colony, and the age of the queen. An old queen will have a lower egg-laying rate and may not be able to lay quickly enough for the nucleus to expand rapidly.

Checking the health of your nucleus

If you have any doubts at all about the health of the brood you should contact your Regional Bee Inspector (RBI), who will advise you. RBIs are employed by the government's National Bee Unit (NBU) to inspect colonies in their region for signs of the statutory notifiable diseases. Each RBI has a team of Seasonal Bee Inspectors (SBIs) who cover their local area. All RBIs and SBIs are experienced beekeepers very willing to offer help and advice on all aspects of beekeeping, not just diseases.

The Inspectorate is a free service to beekeepers in England and Wales. Details are on the NBU's BeeBase website at https://secure.fera.defra.gov.uk/beebase/public/Contacts/contacts.cfm.

Your nucleus may come in a standard nucleus hive, a travelling box or a non-returnable temporary container. Make sure you understand what equipment, if any, needs to be returned or replaced with equivalent new items, for example frames and foundation.

You'll almost certainly have to go and collect your nucleus, especially from a commercial supplier. This is where buying locally helps, as it's good to minimise the stress placed on your colony while being transported.

Transporting bees

For transportation, the inner cover and roof should be replaced by a travelling screen, which is a frame of the same outer dimensions as the box with bee-proof mesh secured over it. The floor, brood/nucleus box and roof must be strapped securely so that they won't move apart in the event of the box suffering an accident. Hive straps are available from equipment suppliers. The entrance must be securely closed. This can be with a metal slide or it can be blocked with a suitable length of foam 'rubber', as found in cushion squabs.

Left: Securing a hive with a ventilation screen for transport.

Left: Blocking the entrance

Make sure the box is properly supported in your vehicle and cannot be dislodged. During transport you need to keep your bees cool, so don't leave them in a hot car or in full sun. To help them keep cool during the journey, you can spray or dribble a little water on to the travelling screen at intervals.

If your nucleus does get too hot, the combs will collapse and fall out of the frames and most of your bees will die. You'll not only have lost your precious colony but you'll also have a dreadful mess to clear up, so take care and look after your purchase.

Transferring your nucleus to a full-sized hive

Unless your nucleus is bursting with bees it can live in its small box for a while, but as it expands it needs to go into a full-sized hive.

At the apiary, place your nucleus in its final position. Open the entrance to let the bees fly and place the roof over the travelling screen to protect from the weather. The bees will orientate and learn the position of their new home. They'll start to fly in ever-increasing circles to become familiar with the hive's general location before setting off to forage over a wider area.

In the next day or two, make sure you have all your hive parts together. You'll need a floor, a brood box, an inner cover and a roof. You'll also need enough frames to fill up the brood box, bearing in mind that you'll be transferring over the frames from the nucleus. If you're a beginner, these will usually be frames fitted with foundation.

Light your smoker and put on your veil. You may also want to put on your gloves, and you'll need your hive tool. (See Chapter 2 for information on protective clothing and beekeeping tools.)

Move the nucleus box and its contents to one side. Put the floor and brood box in its place with the entrance facing the same direction. Don't put any frames in place yet. Gently smoke the nucleus entrance to calm the bees. With a small colony, especially with docile bees, this is probably not necessary but it will do no harm and will give you added confidence.

Left: Smoking the entrance.

Remove the roof and the inner cover from the nucleus. If you feel it's necessary, apply a small amount of smoke over the top bars. Using your hive tool, gently raise first the left lug (if you're right-handed) and then the right of the outside frame against the wall of the box. If you find it easier, especially if you're using a hive that takes frames with short lugs, take hold of each end of the top bar, inside the joint with the side bars.

Left: Lifting the first frame with a standard hive tool.

Left: Holding the top bar to lift the frame.

Removing the first frame can be awkward and you can gain a little extra manoeuvring room by sliding the broad end of the hive tool between the side bar of the outside frame and the hive wall and pushing the frames together. Gently pull the frame out vertically, trying not to roll or crush any bees. Bees are pretty resilient and will survive a certain amount of rolling as the frame is removed, but the less this happens the better.

Below: Pushing the frames away from the hive wall to make frame removal easier.

Transfer the frame and all its adhering bees to the front of the brood box if you want to have the frames the 'warm way' (frames parallel to the front wall), or towards one side if the frames are to be placed 'cold way' (frames at 90° to the entrance). Then transfer across the rest of the frames and their bees in a similar way. It's very important that you keep the frames not only in the same order but with the same faces adjacent to each other. Comb surfaces are not absolutely level and the bees will have adjusted adjacent surfaces to give an appropriate gap (two bee spaces in the brood nest) between them along their length.

Above: Transferring the combs from the nucleus.

Left: Moving two combs at once.

Below: The frames from the nucleus must be in the same relative position when moved to the full brood box.

When all the frames from the nucleus are in the brood box, fill up the box with the spare frames. Place them behind the nucleus frames in a warm-way hive. In a cold-way hive, move your nucleus to the middle, just behind the entrance, and fill up the gaps either side. When your brood box is full, push the frames up together so that the spacing is correct. With Hoffmann frames this can be done in one go from one side of the hive. If you're using spacers, then make sure that each frame is pushed up tight against the adjacent one.

Above: Adding extra frames to fill the brood box.

Above: The brood box with its full complement of frames.

Above: Shaking in the remaining bees.

Above: The faces of the combs are parallel if not flat within the frames.

Add the inner cover and the roof and you have your first colony of bees.

Above: Putting on the inner cover.

Above: Adding the roof to complete the hive.

Collecting a swarm

You can also acquire bees by collecting a swarm. Swarming is the way that colonies reproduce and a swarm leaving a colony is looking for a suitable cavity in which to establish its new nest.

Before the swarm issues from the hive, scout bees have been out looking for new nest sites. The swarm leaves and then clusters not too far from its original location. This may be high in a tree, in a low hedge, on a fence post or anywhere else the queen decides to settle. At this point, scouts 'promote' the site that they've found. Eventually, the colony 'decides' which it prefers and the swarm flies there to take up residence.

Below: A swarm cluster.

Beekeepers are generally called to swarms that are clustering and this is by far the best time to collect them, as once they've gone into a cavity they can be very difficult to remove. For instance, if the swarm has gone into a cavity wall through an air brick, the wall will have to be dismantled to get the bees out. In many cases such measures are impractical and/or uneconomic and the only sensible option is to kill the colony.

Collecting swarms can be great fun but also very frustrating. They can be easy to reach or they can be spread out and present a real problem.

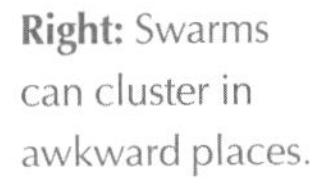

Right: Swarms can cluster in awkward places.

Equipment for collecting a swarm

This is a basic list of equipment needed to collect a swarm. You'll no doubt add to it with experience. If you're prepared to collect swarms, have this equipment ready for when the call comes.

- A strong container, such as a sturdy cardboard box or a skep, in which to collect the bees.
- A piece of loose-weave material such as an old sheet.
- A small block of wood to lift up the edge of the container.
- Strong string.
- A pair of secateurs or loppers.
- A goose wing or bee brush.
- A water spray.
- A smoker, fuel and matches.
- A veil and overalls.

Skeps are described in the next chapter. They're very useful for swarm collection as they're strong and relatively light. The rough inner surface also makes it easier for the bees to cling on. However, swarms can be collected in almost any container. You can make a swarm box out of an unused brood or nucleus box. The floor must be firmly attached and the box should have a ventilated cover and a means of closing the entrance. Swarm boxes can be heavy, which makes them more difficult to use if the swarm is hanging some distance from the ground. A cardboard box must be strong enough not to collapse under the weight of the swarm, which can be up to 4–5kg (10–11lb).

The swarm you collect will generally be some way from your house or apiary, which means you'll have to transport the bees back. The same criteria apply here as to transporting your nucleus hive. The open-weave cloth will be used to cover the open side of the container so the weave must be close enough to prevent the bees from escaping. The cloth must cover the opening with enough extra material to go up the sides of the container. You must be able to secure it over the container with string, baler twine or even a tight elastic bungee cord.

If you're called to a swarm, try to get there as soon as possible. Swarms have a nasty habit of taking off for their new home a few minutes before you arrive!

Below: Securing the cloth over the container.

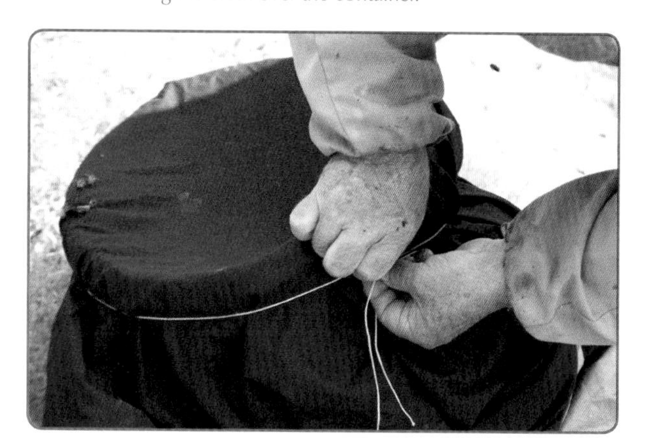

The theory of collecting a swarm

To catch a swarm, you need to transfer the queen and as many bees as possible into your container in one go. Catching the queen is essential to an easier operation, as the swarm bees want to be with her. As she's likely to be in the middle of the cluster, if you can get the majority of bees into your box you're very likely to include her.

Below: A cooperative swarm.

Above: Collecting the swarm.

Above: The swarm in the skep.

The classic swarm hangs in a neat cluster from a branch at about shoulder height. First spread out your cloth near the base of the tree, ready to receive your box. Then lift the container so that it surrounds the swarm and knock all the bees into it with a sharp blow to the branch. Shake the box to knock the bees into the bottom, then hold it over the cloth, turn it over carefully and place it on the ground. Prop up one edge to make an entrance.

Above and below: Propping the skep to make an entrance for the bees.

Lots of confused bees will be flying around looking for the swarm. Bees around the container will start sending out a pheromone message to attract the flying bees. This comes from the Nasonov gland at the end of the abdomen, and a worker bee will bend down the last segment to expose it, then fan her wings to distribute the scent. This is detected by bees in the air and they come down to join their nest mates. However, bees produce pheromones from other parts of their bodies too, one of which is their feet, and this footprint pheromone on the place where the swarm hung will continue to attract bees back there, so you'll probably need to shake the branch again or even smoke the area to disguise the smell.

Above: The Nasonov gland.

Above: Be safe when collecting a swarm.

You should see bees around your container fanning and others landing and marching inside. This almost certainly indicates that the queen is there and, in time, all the swarm bees will join her. You may have to wait until evening when the bees have stopped flying before you can remove the support, wrap the container in the cloth, secure it to prevent any bees escaping and take it to your apiary. During the journey, wedge the box to prevent it falling over. Put the covered opening upwards to give the bees ventilation.

Swarms in awkward places

Unfortunately swarms do not always land in the ideal place. However, wherever they are, the aim is still to get the queen and all the bees into the container. You'll have to assess each situation and work out the best way of doing this.

If the bees are very high up you'll need a ladder. You must make sure that it's steady and safe to climb. It must also take you within easy reach of the swarm. You'll be surprised just how much force is exerted by a swarm of bees falling suddenly into a box that you're trying to support with one hand and just how heavy some swarms can be.

Bees on the ground are fairly easy to deal with because you can put the container over them and they'll crawl up into it. Similarly, if they're spread over a hedge, if you can balance the container above them they'll move up into it. You can encourage them into the container with your smoker, puffing a little smoke behind the group to move it forward. When they're spread up a gatepost or a wall you can scrape them down into the container using something like a goose wing or a handful of grass. Once the queen is in the container, the rest of the bees will generally follow.

If a swarm has gone down a chimney or into a cavity wall, it's best not to try to collect it. Removing them generally requires either a cherry picker to reach the top of the chimney or a builder to remove part of the brickwork. Neither is recommended, especially for a beginner. There's not only a cost involved but it can also be dangerous, and you're unlikely to be covered by your insurance.

Right: This is probably an afterswarm or cast.

Hiving a swarm

It's best to hive a swarm towards the evening as the bees are more likely to stay in the hive. Hive them in the middle of the day and they may decide to look for a different home.

If you've collected the swarm during the day, place a board that is larger than the container next to your hive. Turn your container over and place it on the board. Prop up the edge to give the bees an entrance and let them fly until evening when you can transfer them to the hive.

You can hive a swarm in two ways. You can make them run up a board sloping up to the entrance or shake the bees directly into the brood box.

Above: Flying bees joining the rest of the swarm in the skep.

Running in a swarm

This is perhaps the most exciting way of hiving a swarm. Place your hive in its final position and remove the entrance block to give the bees the maximum entrance space. Find a board at least as wide as your hive and long enough to be propped up to the entrance at a slope of about 30°. Cover the board with an old sheet or other cloth that hangs down over each side, ideally to the ground. This stops any stray bees, especially the queen, from finding their way under the board and clustering there.

Left: Giving the bees a board to run up.

Above: The cloth prevents the bees collecting under the board.

In the early evening, when the bees have stopped flying, invert the container with the swarm and give it a sharp shake over the board. Knock it to make sure all the bees have been shaken out.

Bees will naturally climb upwards and they're also looking for a dark cavity so they'll start walking up to the hive. If you're lucky, you may see the queen running in over the backs of the workers. Once you know she's inside you can be confident the rest of the colony will follow. A little smoke applied behind the stragglers will encourage them to go into the hive.

Left: A swarm running into the hive.

Below: Encouraging the stragglers with a little smoke.

Shaking in a swarm

If you haven't got the time to wait for the bees to run into the hive, you can shake them in.

Prepare your hive in its final location as before but this time remove the brood frames.

Towards evening, take the container with the swarm and shake it sharply, close over the brood box, aiming to get most of the bees inside the hive. Brush any remaining bees out of the container. Now place the frames gently into the brood box. At first they'll balance on top of the mass of bees but as they move up onto them, the frames will sink or can be gently pushed into place. With the frames in place, add the inner cover and the roof and replace the entrance block.

You can shake large swarms into an eke and then place the brood box on top directly. Remember to remove the queen excluder and eke no more than 24 hours later.

Left: Shaking the swarm into an eke.

Left: Adding the brood box with frames and inner cover.

Left: Adding the roof.

Preventing a swarm from leaving

Sometimes a swarm decides it doesn't want to stay in the hive you've so kindly provided and you really don't want this to happen. As a swarm won't leave a hive without its queen, you can prevent the bees doing so by placing a queen excluder (see next chapter) between the floor and the brood box.

This is easy if you shake the swarm into the hive, but if you run the swarm in don't put the queen excluder in place until all the bees are inside. Doing it earlier will prevent the queen from entering the hive – not quite what you want at all!

Remember to remove the queen excluder 24 hours later.

Below: A queen excluder above the entrance prevents the queen leaving with the swarm.

Inspecting a colony

An unobservant beekeeper can open a colony, lift out the frames and put them back having learned nothing. However, the object of looking into a hive is to find out what is going on. At first, as a new beekeeper, you'll have no experience, so every manipulation will be another big step on the learning curve. If you want to learn more quickly, I would urge you to take advantage of every opportunity that presents itself to observe experienced beekeepers handling bees. It's just as important to observe beekeepers who are good at this and those who aren't. You can learn from both.

What are you looking for?

- Does the colony have a queen and is she laying normally?
- Is the brood pattern as expected for the time in the season?
- Is the colony developing as expected?
- Does the queen have sufficient space to lay more eggs?
- Are there occupied queen cells?
- Does the colony have sufficient room for honey storage?
- Does it have enough stores to last until the next inspection?
- Are there any signs of disease?
- What is the colony's varroa status?

Before you open a colony, make sure that you have all the equipment handy that you're likely to need. You can put them in the pockets of your bee suit or in your bee box. Whatever you do, you'll need a smoker and a hive tool.

Below: Check whether the brood is healthy.

Above: The worker's sting.

Getting stung

Before we discuss inspecting a colony, it's worth talking about being stung. It will happen, however quiet your bees and however careful you are. And it will hurt. You'll swell, although this usually reduces in inverse proportion to the number of stings you receive. At first, as the swelling goes down, it's likely to itch, but this will pass. Obviously you'll not go out of your way to get stung, but bear in mind that you're undertaking an activity that increases the likelihood of this happening. If this causes you real problems, then reconsider whether beekeeping is the hobby for you.
If you are stung, the important thing is to remove the sting as quickly as possible, for example, with your fingernail or the edge of the hive tool. If a bee manages to get inside your veil, move well away from the hive before removing it. Alternatively, squash the bee against the veil and kill it.

If you experience a rash, palpitations or shortness of breath, get medical help immediately. Take your mobile phone with you to the apiary for such emergencies, especially to out-apiaries or if you are beekeeping on your own.

Occasionally, people are extremely allergic to bee stings, leading to an anaphylactic shock which can prove fatal. However, this is very unusual, with only 2–3 people dying from bee stings in the UK each year.

If you have experienced a strong reaction to a sting, you should check with your doctor. He may prescribe an Epipen which is used to deliver a rapid dose of adrenalin to counter the reaction. He may also prescribe a course of antisensitisation injections. Taking an anti-histamine tablet a couple of hours before going to the apiary will also help to reduce any reaction if you do get stung.

Making an inspection

Before you go to your hives, put on your veil, overalls, boots and gloves. Make sure you have your hive tool and smoker to hand. Light your smoker and make sure that it's going properly.

The first thing to do is to blow two or three good puffs of smoke into the entrance. Smoke repels bees so this drives away the guards, the bees pre-programmed for colony defence. The second effect is to induce them to engorge on their honey stores and start to fill their honey crops. This doesn't happen instantly. Bees need a minute or two to fill up. Sometimes, bees have no open honey stores so they have to break open sealed cells in order to drink. If your colony is starving, there will be no honey at all and smoking them will have absolutely no effect. If you decide this is the case, close up the hive and give the bees a feed of sugar syrup. You can then inspect them a day or two later. Some may advise you to spray them with sugar syrup. I have tried this: everything becomes sticky and very messy. I wonder if those who give this advice have ever actually tried it?

Opening the hive

Now you're ready to open the hive. Put the smoker down within reach so that it's ready for use when required. What you never put down is your hive tool. You must develop the technique of holding it in your palm with your third and little fingers, using the thumb and first two fingers to lift the frames, etc. After a while it will become second nature and the hive tool will be part of your hand.

Left: Keeping hold of the hive tool at all times.

Lift off the roof and place it, upside down, within easy reach. This is why you need space around your hives. Stand so that your hands fall naturally to the frame lugs; at the side if you have your frames cold-way or at the back for warm-way.

Above: Smoking the entrance.

Left: Guard bees on the alert.

Above: Removing the roof.

Left: The hands should fall naturally to the frame lugs.

Push the flat straight edge of the hive tool between the inner cover and the box and apply a little smoke through the gap. You may need to lever it in several places to break the propolis seal. Removing the inner cover using little or no smoke at all is the ideal. Examine the underside in case the queen is there. She could also be on the underside of the queen excluder if there's one on the hive.

I'm assuming that for your first inspection there are no supers on the hive. Place the inner cover flat, sticky side (underside) up, at an angle on the hive roof. If there are supers on the hive, you don't need to take off the inner cover. Lift it with the top super and stand them on the roof, skewed a little so that the super balances on four small points of contact. Then lift off any other supers one by one and place them at an angle on top of the previous one. In this way, each box only contacts the other where the edges cross, reducing the places where bees can be crushed. If you want to examine the supers, do this as you return them to their place on the hive.

Below: Using the hive tool to separate the boxes.

Above: When separating boxes, some frames may be stuck with propolis.

Below: Using the upturned roof to support the supers.

Having reached the brood box, don't instantly puff smoke across the frames, but pause for a moment or two and look. The frames are there, parallel to you, with bees on and between them. While you pause you give them a chance to settle down, and you'll see that there is a roughly spherical area of the hive where bees are in the greatest concentration. This is centred on where the winter cluster was. This is very evident early in the year, less so later, but every colony expands from and shrinks back to a central area. This may not be in the same place in the autumn as it was in the spring, but all colonies have this centre.

Below: Bees over the top bars.

Removing the first frame

You need to loosen the first frame before removing it. If you're right-handed, always start to loosen on the left, then move to the right and vice versa. Some bees react to movement but by working this way you cut down the number of times your hands move across the bees. If you have a 'J' hive tool, put the hook end under the lug of the first frame and sit the square end behind it on the next frame. Push away from you at the top. This levers the frame upwards and you can grasp the lug with your left hand. Then move to the other side. Once the frame is loose enough, you can lift it upwards. With a standard hive tool, carry out this manoeuvre using the curved end. With Hoffman frames you can use the flat end of the hive tool to first push the frame's side bars away from the outside wall and then back from the second frame.

Above: Lifting the frame lug with the 'J' hive tool.

Above: Loosening the first frame.

Lift the frame slowly and smoothly, trying not to jerk or bang anything. There is not much spare room and there will be bees between the comb and the hive wall and the adjacent comb, so try not to roll them around. If you move gently they'll be able to cope and walk away when you've finished.

This first frame may be full of honey and covered in bees or it may be empty. Look carefully and try to identify as many things as possible. Is there any brood present? If so, is it eggs, larvae or sealed brood? Does it contain stored pollen? How much honey is there? Hold the frame over the hive while you examine it just in case anything vital, such as the queen, falls off.

Put this frame to one side. You can hang it in a spare brood or nucleus box or place it in a cardboard box. If your hive stands on rails, you can put one lug on the far rail with the bottom bar resting on the near rail. You can also lean it against the hive stand at the front of the hive on one lug and a bottom corner.

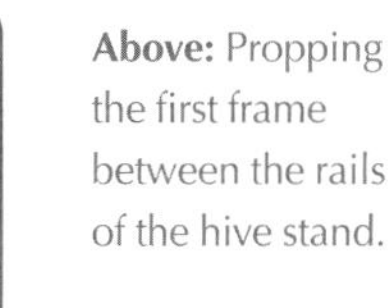

Above: Propping the first frame between the rails of the hive stand.

Left: The first frame propped at the front of the hive.

Turning a frame

If there is fresh nectar in the cells, it may well fall out as you tilt the frame so you need to learn to reverse it in a vertical plane.

1 Hold the frame by the lugs.

2 Drop one hand down and raise the other so that the top bar is vertical.

3 Rotate the frame through 180° in the vertical plane so that the other face of the comb is towards you.

4 Raise the hand you dropped and drop the hand you raised. You'll now be looking at the other side but with the top bar at the bottom.

Turn it back in the same way ready to be replaced in the hive.

Creating a space

You have now created a gap. As you take out each frame in succession, it's replaced on the side of the gap nearest to you and moved firmly back towards you. The object is to retain the gap next to the frame you plan to remove next. As you continue to inspect the frames, the gap moves with you.

Above: The working gap which moves with you across the brood box.

You may need to apply a little smoke to move the bees away from the lugs of the next frame. To remove it and each subsequent frame, start on the left (or the right if you're left-handed) and lever the frame away from the others into the gap. Use your thumb to stop it jerking as it moves. Lift it out smoothly to eye level for inspection. If you can't see the comb you need to remove the bees. Hold the frame by the lugs down in the gap and give it a sharp shake. This will dislodge most of the bees into the hive. Look for the same things as you did on the first frame. If you're close to the brood nest you should find a lot of pollen, particularly on the side adjacent to the brood. The different colours will reflect the variety of flowers your bees are working.

Right: Breaking the seal between frames.

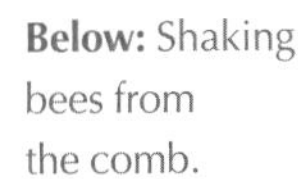

Below: Shaking bees from the comb.

Above: Pollen of many colours indicating which flowers your bees are visiting.

If the colony is expanding, the next frame may well have an area of 'empty' cells. This is where you need to be able to see the eggs the queen has laid. Seeing eggs tells you that something is laying in the colony, whether it be a normal queen, a drone-laying queen or laying workers. You can confirm which when you reach the larvae and sealed brood.

When you've finished your inspection, replace the frame into the gap, making sure that it's facing in its original direction. Combs are not necessarily completely flat and the parallel contours of adjacent surfaces need to be kept together. Move it firmly to your side of the gap so that this doesn't become smaller as you move across the hive. You can then remove the next frames in turn and inspect them in the same way.

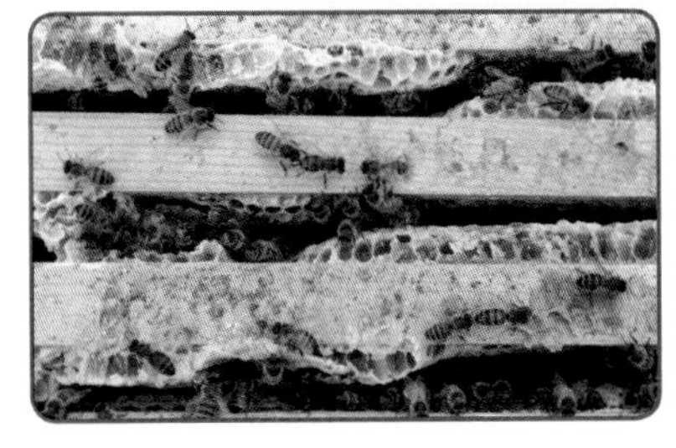

Above: 'Empty' cells may well contain eggs.

Left: Wavy comb means frames must be kept in the same order.

The brood nest

As you remove each comb you'll be looking at a slice of the spherical brood area which is encased in a 'shell' of pollen. This is thicker at the back, thinner at the top and sides and very thin, if present at all, beneath the brood. On cold-way frames the brood area will be towards the side of the comb nearest the entrance. When you reach the brood nest, you'll see that the patches of brood increase in size to a maximum and then decrease on the other side. The centre of the brood nest is not necessarily in the middle of the hive but it's above this point that bees are most likely to start storing honey in the super.

You'll see a variation in the age of the brood. Eggs will be next to small larvae. The larvae will get bigger the further your eye moves away from the eggs and a similar graduation in development goes on unseen under the cappings. Eggs take three days to hatch. Worker larvae are sealed approximately six days later. Worker brood is sealed for 12 days, giving a development time of 21 days. It also gives you proportions of 1:2:4 for the three stages.

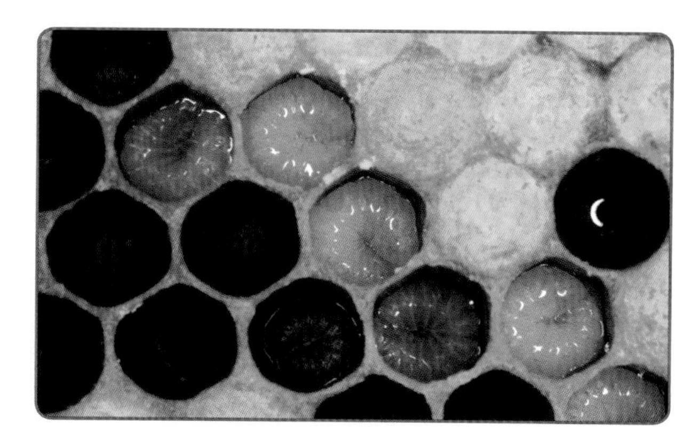

Above: Eggs and young larvae.

Above: Older larvae and sealed worker brood.

If your colony has reached its peak size, those are the proportions you'll have. If it's expanding, there will be a higher proportion of eggs and larvae. When it's contracting, more sealed brood will be evident. If you have few eggs and unsealed queen cells, your colony is trying to swarm. If you have no eggs and sealed queen cells, your colony has definitely swarmed.

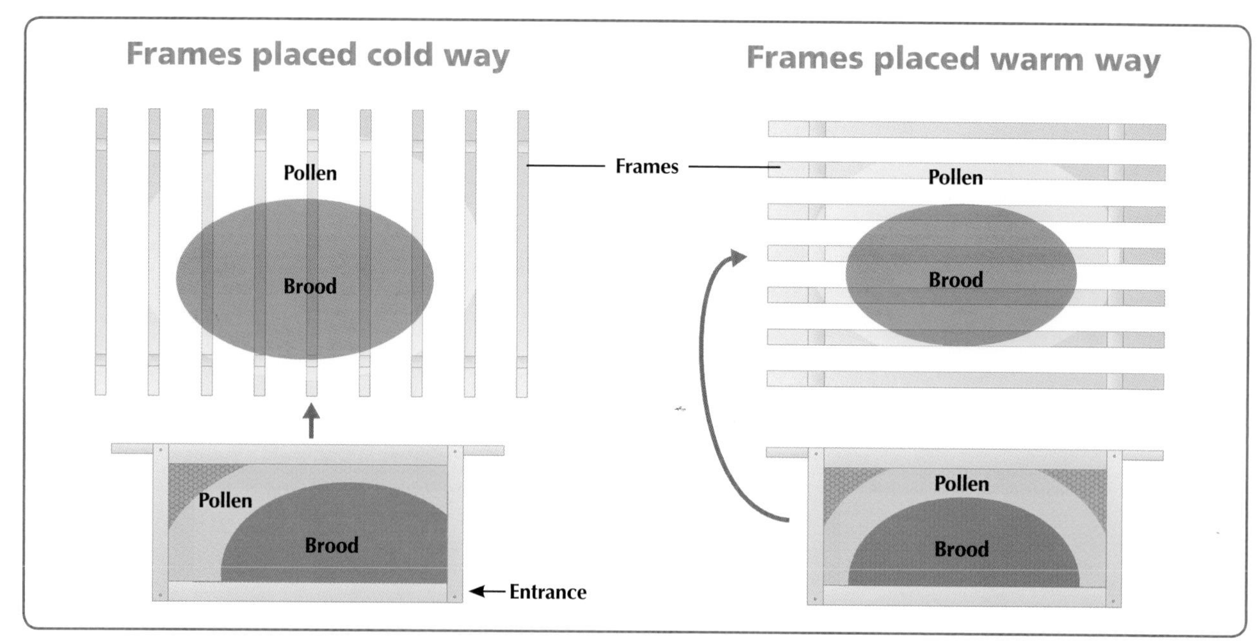

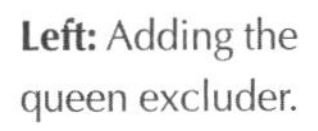

When you've finished your inspection, the gap will be at the other side of the hive. You need to move the frames back to their original positions and replace the first one. If you're using Hoffman frames they can be moved en bloc with your hive tool. Push the flat edge between the hive wall and the side bar on one side, then the other and move the frames over. You may be able to achieve this either in one go or in small groups. Hoffman frames maintain the spacing and you should not crush any bees, with the risk of spreading disease. If there are lots of bees on the back hive wall, persuade them to move with a little smoke. With DN1 frames and spacers you'll have to move each frame in turn. Press the frames firmly back into their original positions.

The gap is now back where it started. Picking up the first frame with the same hand you used to put it down will automatically put it back in the same relationship to the others.

Your first inspection needs to be at least six to eight weeks after breeding starts in the colony. Early to mid April is soon enough but this will depend on your geographical location. If it's warm enough for you to go out in the garden in short sleeves, then it's warm enough to open your colony. However, particularly early in the year, don't have the colony open for too long as this might cause the bees to kill the queen. Don't rush. Take things calmly and steadily.

The whole point of regular colony inspections is to find queen cells before they're sealed. You do not need to start regular, seven- to nine-day inspections in earnest until the colony is producing drones.

Closing up

When you've replaced the last frame you can begin to close up the hive. If the bees are really filling the brood box and building bits of comb on the top bars, scrape these off with the flat blade of the hive tool and collect them in a covered container. Then put a queen excluder on the hive, followed by a super. Queen excluders framed on one side only go on the hive frame-side down if the hive has a bottom bee-space, or frame-side up for top bee-space. Finally, replace the inner cover the way it was originally, and then the roof.

Below: Scraping burr comb from the top bar.

Left: Adding the queen excluder.

Left: A framed wire queen excluder.

When your colony has supers on you should not need to inspect individual frames, as the weight of the box will give you an idea how full it is. You'll be able to see if the bees have capped the cells by looking between the frames but when you're just starting you might find it helpful to have a quick look.

Keeping records

You might have closed up the hive but your inspection is not yet over. You need to record what you've seen and done, as was discussed in the previous chapter. It really is important to get into the habit of keeping records. The best memory in the world sometimes fails and that might just be at a vital time in your swarm control or queen-rearing programme. It's best to spend a few minutes making notes before you move on to the next colony.

Below: Keeping records is important.

Finding the queen

Most of the time you only need to be able to detect the presence of a queen that's laying normally. However, there are crucial moments when it's most important to be able to find the queen herself, such as when the colony is trying to swarm, when you want to dispose of her, or when you want to mark her. You're much more likely to succeed when the colony population is low, which is usually early in the year or when a newly reared queen is just starting to lay in her nucleus.

Marking the queen

Beekeepers use a colour code to indicate the year a queen was hatched. Five colours represent the final digit of two years, five years apart. I use a mnemonic to help me remember which is which:

Year ending	Queen colour	Mnemonic
1 or 6	White	What!
2 or 7	Yellow	You
3 or 8	Red	Rear
4 or 9	Green	Green
5 or 0	Blue	Bees?

Above: A marked queen.

Queen-marking paint is available with which to mark her on the thorax. If you want to number your queens, you can buy kits with small numbered discs to stick on. If you're keeping records, you don't have to follow the code and you can use the colours that you can see most easily and then just record this. I find yellow and white show up best on black bees with blue and green easier to see on yellow bees. Look at the available colours and choose what you think will suit you.

Above: A numbered queen.

Have your marking paint ready with the lid on loosely. The brush in commercially available paint is much too big. I use something like a matchstick. Dip it into the paint on the brush or in the pot and then apply a small dot rather than a big splodge.

Open your colony using as little smoke as possible. Queens are shy of bright light and she may well be on the other side of the comb that you're looking at. It's also worth looking down into the hive at the face of the next comb to see if she's there.

You're looking for something different. The queen is not that much bigger than the workers but she does have distinguishing features. I look for her long legs but others will look for her longer abdomen. Take out the frame and hold it over the hive. Have a picture of a queen in your mind and scan down one side of the patch of bees on the comb. Then look along the bottom, up the other side and along the top. Finally look to and fro across the centre. The queen could well move swiftly round the back of the frame and you may spot her as she goes over the edge. Turn the comb over and check the other side.

Below: Looking round a frame for the queen.

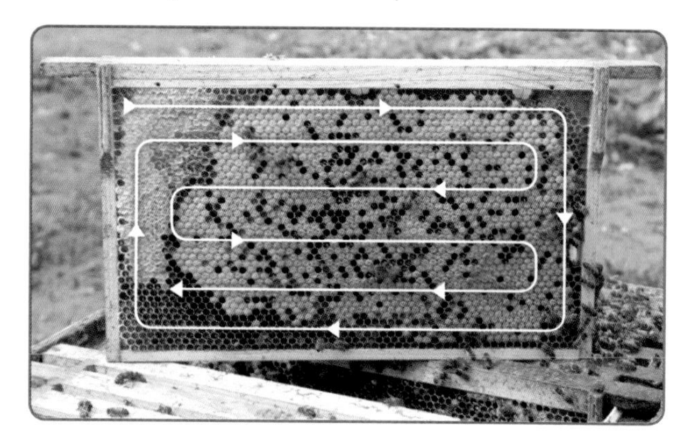

Take time to look properly but don't take too long, because if the queen isn't there you won't find her. The longer you spend looking at one comb, the further into the brood nest she could move. Repeat the process with the rest of the frames. If you haven't used very much smoke, pay particular attention to combs with large patches of eggs, as she may be there. Don't forget to look at the side bars and the bottom bars.

When you find her you need to restrain her, and I would recommend what is known as a Baldock, 'crown of thorns' or press-in queen cage. Keep your eye on the queen and when she's on the flat face of the comb, place the cage over her and press down gently! The spikes will hold the cage in the comb and they're set too close for the queen to escape. Ease up the cage slightly so that she can just move around. The aim is to get her thorax in a square of the threads across the cage. When she's in position, press down just enough to prevent her from moving. Take your matchstick and apply a small amount of paint to her thorax. Wait a few seconds for it to dry and then lift the cage off with a smooth vertical movement. The queen will move as soon as she can and you don't want your neat blob smeared. It's worth watching her until you know that she has been welcomed back by the bees. Some workers take exception to the smell of the paint and can attack the queen. If this happens, break up the ball of bees surrounding her and hopefully all will be well.

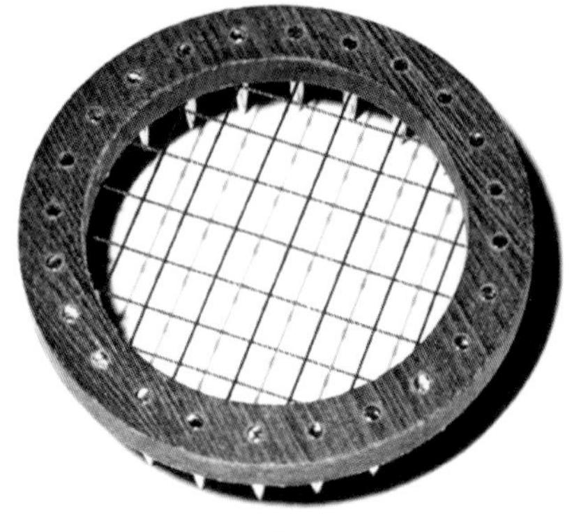

Left: A Baldock or 'crown of thorns' queen cage (upper surface).

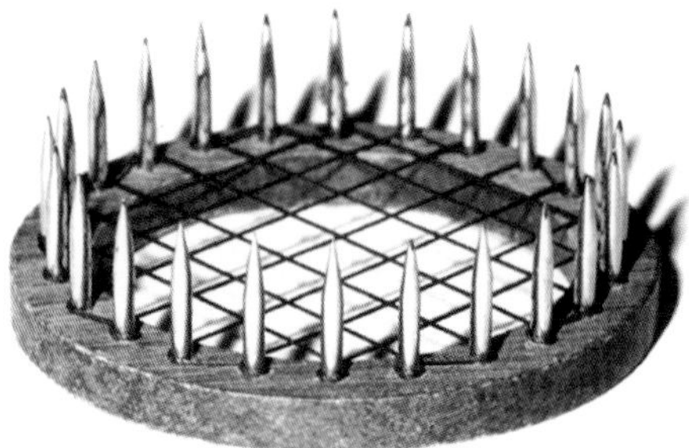

Left: Lower surface.

You can pick up the queen to mark her. If you are right-handed, hold her by her thorax or wings with your right hand and offer her the second finger of your left hand to stand on. Let your left thumb and first finger relax and hold her thorax. Then apply the paint. Reverse the directions if you are left-handed.

Above: Holding the queen by her wings.

Above: Offering her a finger to stand on.

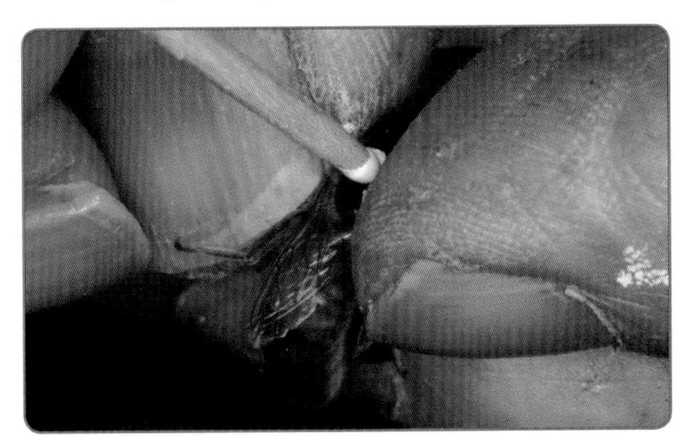

Above: Marking the queen's thorax.

Before you mark your queen, you might like to practice on drones as you can pick them up with no danger of being stung and if you do happen to damage them, they are dispensable. However, either kill any drone you mark or use a different colour from that for the queen or you will find it very confusing when looking for the queen in your colony.

If you don't find the queen on the second pass through the brood nest, close up the hive and come back another day. If you get really stuck, ask an experienced beekeeper who marks his queens to help. They're usually very pleased to do so.

Swarming

A colony of bees that survives and prospers may well eventually decide to swarm, which is effectively colony reproduction.

To the casual observer it's as if the bees have gone mad. Suddenly, prompted by nothing the observer can see, they start to pour from their entrance. They swirl about in the air, flying in ever-increasing circles. Eventually about half the bees in the hive will be in the air. The combined buzz can be loud and alarming – and to the beekeeper, very exciting!

Above: A swarm in the air.

Above: A 'classic' swarm cluster.

The cluster

As you watch, the bees will start to form a cluster. This can be on anything – a wall, a tree, a wheelbarrow handle or a car bumper. I once knew a lady who had a swarm come through her bedroom window and cluster on the foot of her bed! If they cluster in an accessible place you can collect the swarm, as already described in Chapter 3.

The cluster forms where the queen has landed. The dark ball of bees attracts more bees. Sometimes the queen lands and then decides to fly off again. At each of these stops, she leaves a trace of 'queen substance'. Small temporary clusters may form at these points because they're attracted to the queen's pheromone, which sends out the message that she's present. Of course, she isn't, because she's moved on, but her scent is still there.

Queen substance is a complicated mixture of many fatty acids that, together, act as a pheromone, affecting the physiology and behaviour of the other bees in the hive. Two components are more important than the others. In the case of the swarm, one part attracts the bees to move with the queen in the air while another helps to stabilise the cluster and keep it together.

Once the cluster has formed it can stay very still and quiet, so that in the right surroundings swarms can actually be very hard to spot! The cluster usually forms within 10–50m (30–160ft) of the original hive. However, some bees will fly a long way from the hive before clustering. The vast majority of swarm clusters will be found within a radius of some 200m (650ft) of their original nest site.

Finding a new home

The swarm is looking for a new home. Even before it issues from the hive, scout bees have been out looking for a suitable cavity. They go inside and then fly around, effectively measuring its volume and determining if it's large enough for the new nest. Information about the various possibilities is passed to the bees in the cluster by dances which take the same form as the foraging dances described on page 103.

Eventually a decision is made and the swarm departs for the chosen site. This can happen after a very short time or it can take several days. Occasionally, if the swarm hangs in a very sheltered, shady place and the weather turns wet and cold, the bees begin to build comb in the place where they're clustering. However, such swarms rarely survive because they cannot defend their honey stores against robbers. They cannot produce sufficient bees to fight off the robbers, keep the brood nest warm or collect enough food to last them through the winter.

The beekeeper can collect a swarm most easily when it's clustering. As described earlier, a suitable container can be placed around the bees, which are knocked into it and then taken back to the apiary.

Above: A swarm which has built comb in the open.

Colony reproduction

Bees exist on several levels. They're individuals but they live in a colony which can be regarded as a superorganism, *ie* a group of individuals each of which is unable to live independently and where individuals play specific roles in the survival of the group.

When a colony swarms, where there was one superorganism there are now two. Swarming can therefore be regarded as colony reproduction. Just as animals grow and develop to the point where they can reproduce, so does a bee colony. Being able to inspect the colony gives the beekeeper the opportunity to control the colony reproductive process so that it doesn't swarm and no bees are 'lost'.

In order to control swarming, you must be aware of how a colony that comes through the winter develops in order to reach the point at which it prepares to swarm. This will help you identify when you can intervene with the most effect.

As the day length increases in spring, the colony begins to raise the temperature at the centre of the cluster to around 35°C (95°F), enabling brood rearing to take place. House bees must clean out and polish the worker cells before the queen will lay eggs in them. Bees need energy to raise the temperature and this comes from their honey/sugar stores.

The queen begins to increase the number of eggs she lays and these go through the normal metamorphosis before emerging as adults from the sealed cells. Initially, the number of bees increases slowly, though larger colonies can obviously grow faster than smaller ones. Growth in numbers is offset by the deaths of the adult bees that have lived through the winter ('winter bees'), which are required to work hard to create the heat to increase the colony temperature.

'Winter bees'

'Winter bees' have a different physiology from 'summer bees'. Whereas summer bees will live around five or six weeks, winter bees are able to survive for five to six months, depending on the latitude (in colder, more northerly areas they'll be unable to forage for a longer period than in warmer, southerly areas). In summer bees, the hypopharyngeal glands in the head – which produce brood food for the larvae – are only active for a few days, but those in winter bees will still be active the following spring. However, to produce brood food these bees must have access to pollen as well as honey. Initially the only pollen available will be that collected in the autumn (see page 132). However, as ambient temperatures rise and fresh pollen becomes available from early spring flowers such as pussy willow (Salix), crocus and single snowdrop varieties, larger areas of the comb can be cleaned and warmed and the brood nest expands. This starts slowly but becomes very rapid in the weeks before a swarm is likely to emerge. In general, a colony should be storing honey in the super before it thinks about swarming.

Above: Collecting pollen from pussy willow in spring.

Below: Grape hyacinth.

Below: Single snowdrop.

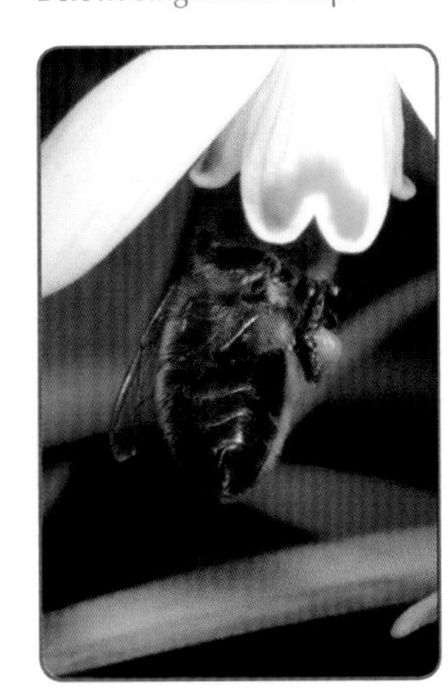

Drones

When colonies are prosperous and thriving in the spring they're willing and able to produce and support drones. No colony without drones is in a position to swarm, so when you see eggs and larvae in drone cells, you know that swarming may occur. Drone brood is not necessarily a sign that the colony is making preparations to swarm, but when there is none you can be certain that it isn't. Colonies will produce drones whether or not they're going to swarm, and removing drone brood will certainly not stop them doing so. Drones are a vital part of colony development and you should not try to prevent the colony from raising them.

In the English Midlands you can find drones in a colony as early as mid-March in a warm spring. Further south, say in Cornwall, they'll appear much earlier, while in the north they'll not be raised until later.

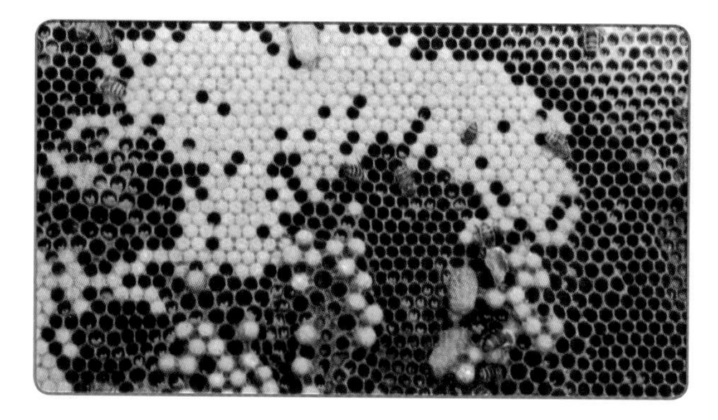

Above: Flat worker and domed drone cappings. Note the sealed queen cell.

Queen cups

The next visible sign that a colony may be considering swarming is the construction of the bases for queen cells, known as queen cups. They look like acorn cups and face downwards from the edge of the comb. They're built around the brood nest, particularly at the lower edges of the combs. Bees also build them on the edges of comb that has been damaged. If you're using a brood-and-a-half or double-brood system (see page 34) and the brood nest is spread over the two boxes, a favourite site for these queen cups is along the bottom edges of the combs in the top box.

Left: A queen cell cup.

Extra room

By this time colonies should have expanded so that they're at least filling the boxes in which they overwintered with bees, and therefore require more room. If you don't give them extra room at this stage they'll almost certainly swarm because they're congested. Work by Professor Thomas Seeley of Cornell University has shown that bee colonies prefer cavities with a volume of around 40 litres. This means that the nest will have become congested by the time natural swarming takes place. In the United Kingdom, this is from mid-April to mid-July. A few precocious colonies have been known to swarm as early as March and some tardy ones to delay until August, but for most areas this is extremely rare.

Queen cells

The colony should continue to develop. Worker bees clean out the queen cups and the queen lays fertilised (female) eggs in some of them. She does this in batches so that the ages of the developing queens can vary. A queen cup becomes a queen cell when it contains an egg. At this point workers will begin to elongate the cell walls vertically downwards, and it becomes obvious which ones contain developing queens.

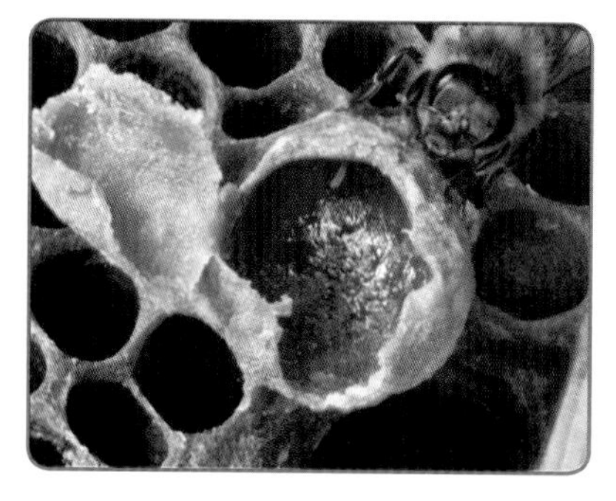

Left: An egg in a queen cell.

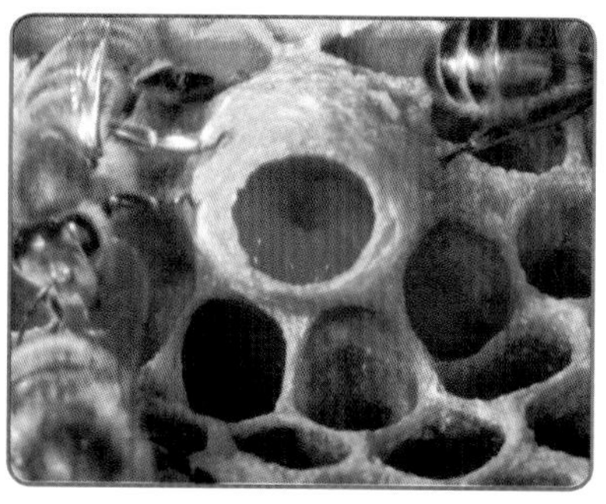

Left: A larva in a queen cell.

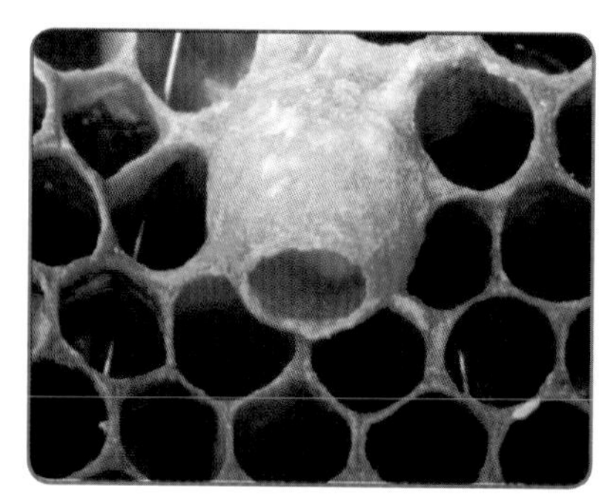

Left: Walls of the queen cell begin to be elongated.

Timing

You need to fix the timing of the developmental stages of queen bees in your mind, as all methods of swarm control involve actions that are linked to stages in this process. Bees can raise new queens from worker larvae less than three days old.

If you're to make proper, effective decisions and undertake appropriate manipulations, you must be able to see and identify eggs in worker cells. If necessary, take the flat end of your hive tool or a sharp knife and cut vertically through a short line of cell walls. Use the blade to bend the walls of the cells, opening them up, making it easier to see into the bottom of the cell. Choose an area of the brood nest where you would expect to find eggs, bearing in mind that the queen will start laying in the centre of an empty comb and work her way outwards in circles. If you find it difficult to see eggs, take the frame into a well-lit area of the apiary and position it so that the light falls into the cells.

Above: Eggs in the brood nest.

The decisions you make about how to control swarming in a colony when there are eggs in worker cells will be very different from those when no eggs are present.

The larvae in the queen cells will develop in relation to the time the eggs were laid. The egg hatches on the third day after it's laid, the cell is sealed on the eighth day, and the virgin queen emerges on the sixteenth day.

Soon after the first queen cell is sealed the swarm is able to leave the colony. However, the colony can change its mind at any time during the process. If it decides not to swarm, workers will tear down the queen cells and life will return to normal.

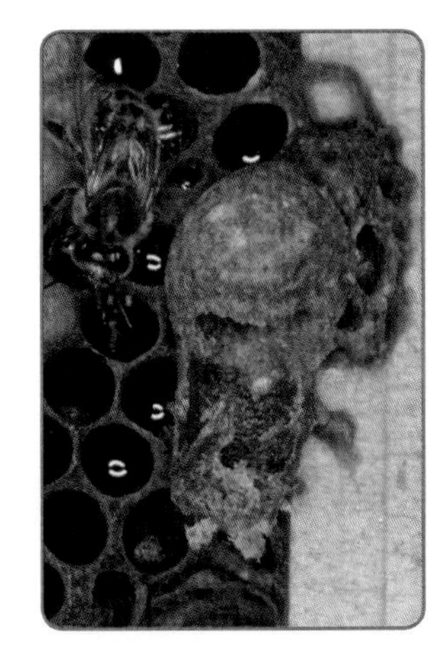

Above: A broken down queen cell.

Summary

These are the stages of colony development to the time it swarms:

1 A healthy colony overwinters and starts to develop and expand.
2 Colony size starts to increase rapidly.
3 Drones are produced.
4 Colony size continues to rise and the bees start to build queen cups.
5 The queen lays eggs in the queen cups in batches.
6 The eggs develop into larvae and the egg that was laid first is the first one to be sealed.
7 The old queen with about half of the bees emerges from the colony on that day or, in bad weather, on the next fine day.
8 About a week after the swarm has left, the virgin queens begin to hatch. Each has the potential to depart with a group of bees in an 'afterswarm' or 'cast'. If several virgins hatch at the same time, they may all leave with the afterswarm. In this case, when the bees have taken up residence in their new home the virgin queens will fight to the death, hopefully leaving one to fly out to mate and head the new colony.

Swarm control measures must be applied between stages 5 and 6 and between stages 7 and 8. See the next chapter for details.

Supersedure

Honey bee queens aren't immortal. Many die in the winter, resulting in the loss of the colony. Some cease to lay worker eggs and these colonies peter out as the workers die. Bees cn detect failure in some queens and will rear a replacement for her without actually swarming. This act of queen replacement is called supersedure. In regard to failing queens, this act is simply that – the replacement of a failing queen. However, some queens live to a good age, say three, four or more years, and are then replaced. In strains of bees where swarming is not a regular occurrence, the supersedure of an old queen is highly significant. Such strains are very valuable to the beekeeper.

Supersedure queen cells are generally large. They are frequently, but not invariably, built on the face of the comb and are usually very few in number (1–3). The process usually occurs in the late summer, say around August. It has been said that the queens from supersedure cells come from defective, failing mothers. This may be so when young queens are superseded but if the mother lived a life of three to four years with good results, how can the daughter queen be anything but desirable? Such long-lived, non-swarming mothers are the basis of a low-swarming strain.

Swarm prevention and swarm control

Some relatively experienced beekeepers see no difference between these two terms and use them indiscriminately. However, for our purposes, I would like to define them.

- *Swarm prevention* comprises those activities that help to delay or entirely stop any swarming preparations before they start.
- *Swarm control* comprises those activities that deal with a colony that has begun swarming preparations.

Many beekeepers seem to have a hang-up about how complicated swarm control is, but right from the beginning I would like to say that it isn't difficult, and it's entirely possible to stop colonies from swarming.

Swarm prevention

This involves giving the colony room beyond its current requirements. When a colony is developing and building up, the bees tend to store incoming nectar away from the brood nest. Once they've filled up this available space and begin filling cells in the brood nest, the area of comb in which the queen can lay eggs is reduced. This seems to be one of the triggers for a feeling of congestion that may initiate swarming preparations.

Nectar contains more water than honey and therefore occupies a greater volume. Giving the colony more storage space reduces congestion and can turn the colony's mind from swarming. It's best to add drawn super combs (see page 61), but as a beginner you may not have access to these. The next best thing is to add frames with 'starter strips' of unwired foundation. Cut strips about 25mm (one inch) wide and fasten one into a frame in the same way as a whole sheet of foundation. Why is this better than giving the bees frames with a full sheet of foundation? Because this natural reaction to a space in a cavity is to build comb to fill it, and this gives them wax-making/comb-building work to do that will help reduce their swarming impulse. However, if you only have frames fitted with foundation they'll be perfectly adequate. Comb built on starter strips can be used for cut comb.

The rule of thumb is this: if the brood box or the super on the colony is full of *bees*, then give them another super. If you can't make up your mind whether or not another super is needed, give it to them anyway. If they don't need the space they'll ignore it. If they do and you fail to provide it you may regret it.

On the other hand, plenty of colonies swarm when they have lots of room. Spare space in a colony when there's no honey flow has little or no effect, but if there's a flow and the bees have room to build comb this can actually help. Colonies busy making wax are less likely to try to swarm.

Left: A super frame with a starter strip of foundation.

Regular examinations

Once you've seen a queen cell you'll realise they're very obvious on the face of the comb. A queen cell that's about to be sealed is about the size of an unshelled peanut with a similar surface texture. You need to be able to see queen cells for your beekeeping to be successful.

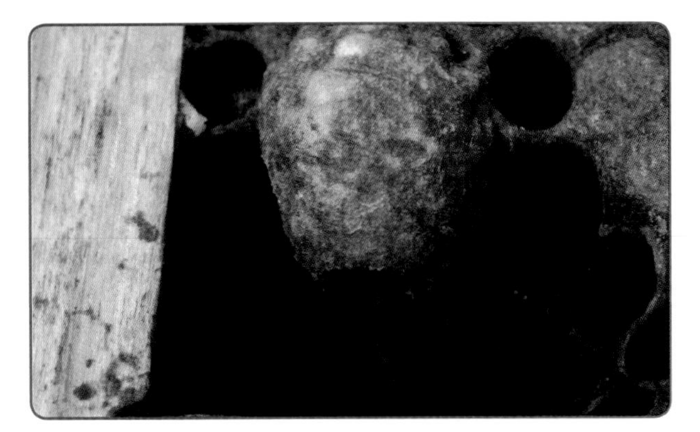

Above: An occupied queen cell.

More importantly, you must be able to see and identify eggs. They're about 1.6mm long and 0.6mm wide and look like tiny white bananas standing on their narrow end in the bottoms of the cells. You must be able to see them in both worker and drone cells and in queen cell cups. If you can't see them on your own, get an experienced beekeeper to show you. Don't give up until you and your mentor are sure that you can see eggs. Put on your reading glasses or use a magnifying glass. Cutting through the walls and opening up the cells in an area where you expect to find eggs can also help.

Above: You must be able to see when eggs are present.

The next stage is to inspect the brood nest regularly, weather permitting, soon after your colony starts producing drones. You're looking to see if the queen has laid eggs in queen cups. Open the colony for inspection, remove the supers and look in the brood box. If there are no occupied queen cups (*ie* none containing an egg), reassemble the hive. As you do so, check if the bees need more room and, if they do, add a super.

How regular is regular?

How often should you inspect your colony? Think of this in terms of the development of a queen. Eggs take three days to hatch. Queen cells are sealed on the ninth day after the egg is laid. This is why beekeepers talk about a nine-day inspection regime. However, this doesn't fit easily into our seven-day week so most beekeepers look every seven days. The shorter time between inspections should actually give you a better chance of finding queen cells before they're sealed.

Sadly, if you find any sealed queen cells it almost certainly means that you failed to spot them on your last inspection and your swarm has left. Another clue that the swarm has gone will be that the number of bees in the colony is significantly reduced. We'll look at what to do then when we consider swarm control.

You should inspect the brood nests of your colonies regularly when you're looking for signs of swarming preparations. However, once you find occupied queen cells future inspections are determined by the method of swarm control you choose to adopt. Once you've sorted the situation out your regular examinations are no longer necessary. If you're fortunate enough to have a strong colony that can collect a good crop of honey and it doesn't try to swarm, you can stop looking regularly around the end of June, depending on your geographical location.

Right: A sealed queen cell.

Swarm control

Although not all beekeepers like to do so, something that certainly helps to control swarming is to clip the queen's wings on one side. This unbalances her and she cannot fly with the swarm to its new nest. Without the queen, the swarm will return to its original hive. It's a bit like belt and braces in that if things go wrong during your swarm control operations, you won't lose 10–14,000 bees. Keeping your colony together, especially during a honey flow, will mean the maximum number of bees will be available for nectar collection and for your honey crop. If the queen tries to fly regardless and gets lost on the ground near the hive, you already have the means to replace her in the form of the queen cells in the colony.

By 'things go wrong' I obviously mean beekeeper error, but natural circumstances can also alter events. For example, in a heatwave dark hives placed in the sun will get hot. This seems to encourage the bees to swarm early to relieve the congestion. It's then not uncommon to find swarms leaving hives before the queen cells are sealed. Even regular seven-day inspections wouldn't help to prevent this, as it's out of your control.

If you decide not to clip the queen's wings, you must be very conscientious in your brood nest examinations and be ready to accept the occasional lost swarm. In my experience, clipped queens live just as long and are just as effective as unclipped ones. You must decide for yourself.

The common denominator

All swarm control methods deal with the same sequence of events, and if you can learn the basic principles you'll soon understand what's being done. All methods are fundamentally the same.

Colonies that are trying to swarm should be thought of as consisting of three parts:

1 The queen.
2 The brood.
3 The flying bees.

All methods of swarm control involve separating one of these from the other two. For example, you can remove the queen and leave the brood and the flying bees or you can take away the brood and leave the queen with the flying bees.

Bees remember their hive entrance as a point in space. If you move a hive 2m (6ft) from its original position, the returning foragers will ignore it and fly around at its previous location. If there's no other hive closer to the original site, they'll eventually find their way home again, 2m away. However, if you have another hive nearer to the original location they'll beg to be let into that one (see Chapter 2).

By moving the hive, containing the brood and the queen, away from its original site you have, in fact, separated the flying bees from the brood and the queen.

The queen and brood can be physically removed from the swarming colony. Marking the queen will help you to find her if you need to but it's possible to deal with a swarming colony without doing so, and we'll consider this later. Removing the brood usually involves physically moving the whole box with frames and bees in it, or maybe transferring the combs one by one, with their adhering bees, into another box.

Below: A clipped and marked queen.

Above: Bees build wild comb if frames are missing from the brood box. These are suspended from the roof of an empty nucleus box.

'Spare' equipment

Most methods of swarm control involve using extra equipment. For instance, in the 'artificial swarm' method you'll need another floor, brood box, frames, inner cover and roof. If all your equipment is full of bees and you have nothing spare, your swarm control will have to start with killing the queen!

All this 'spare' equipment needs frames with foundation or drawn comb. If you make sure that all your spare boxes are equipped with their full complement of frames, you'll always have the right number available for the boxes involved in your chosen swarm control method and you'll never leave gaps in which the bees will build wild comb independent of the frames. As has already been described, bees have an instinct to fill large gaps with comb, which they'll begin on the top surface – be it the inner cover, queen excluder or the bottom bars of frames in the box above. This can produce a tricky situation to sort out but can be totally avoided by making sure that after you've finished your manipulations, all the boxes contain a full set of frames.

'Nucleus' method

This involves removing the queen from the swarming colony and placing her in a nucleus box. She is your insurance if something goes wrong and your new colony fails to raise a queen or she fails to get mated. The queen is thus separated from the brood and the flying bees. If you don't have a nucleus box, you can use a full-sized brood box, but in either case make sure that the box contains a full set of frames.

1 There will be queen cells in various stages of development. If you've been carrying out inspections properly, none will be sealed.

2 Place your nucleus box near to the swarming colony and remove the frames. Find the queen.

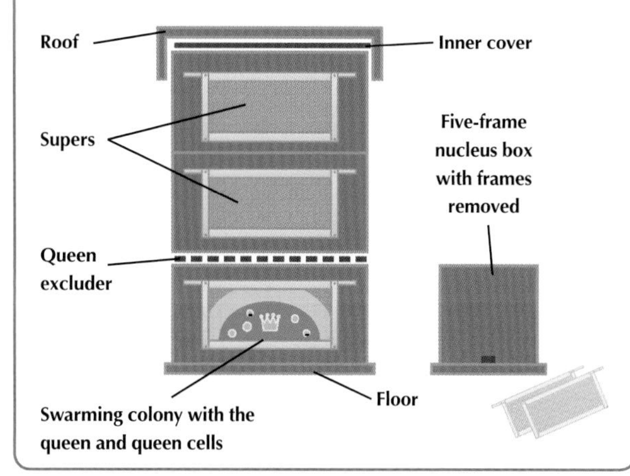

3 Place the queen, on her frame, in the middle of the nucleus box. Destroy any queen cells on this frame.

4 Find two frames in the swarming colony containing a good amount of food and place them on each side of the frame with the queen. Move all three frames to one side of the box.

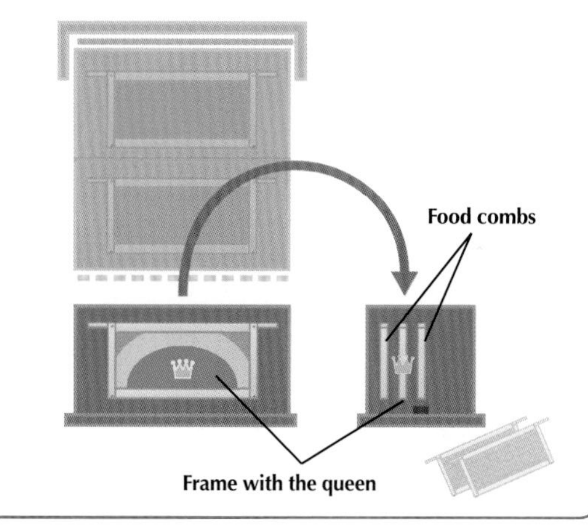

5 Shake the bees from two more frames in the swarming colony into the nucleus.

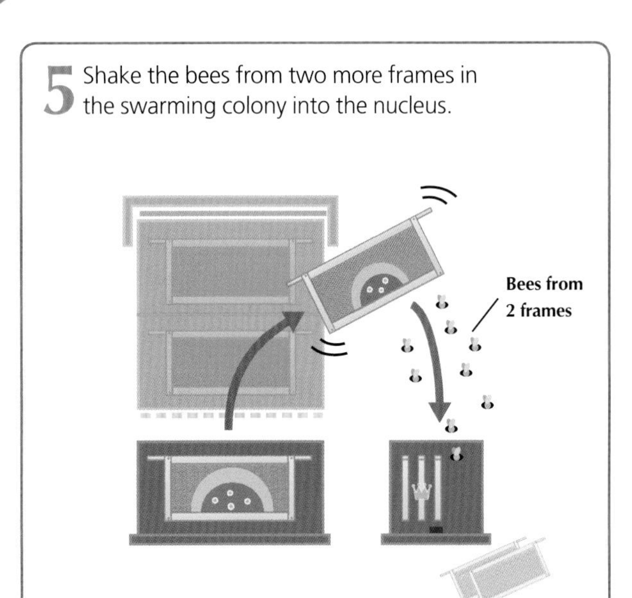

6 Fill up the gap in the nucleus box with two of the spare frames.

7 Stuff grass in the entrance and place the nucleus with the queen on a new stand elsewhere in the apiary.

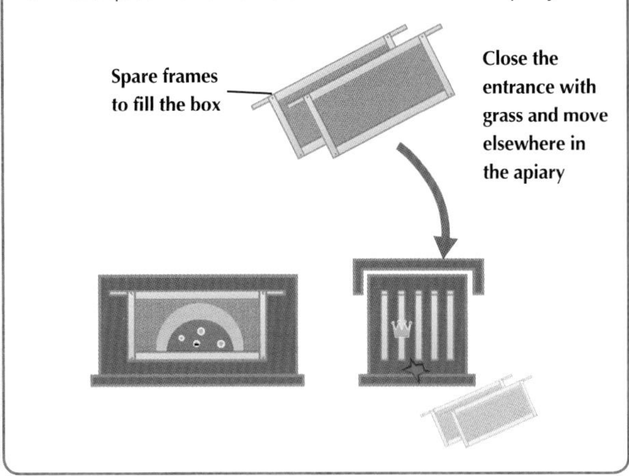

8 Examine all of the remaining frames in the swarming colony. Make sure that they don't contain *any* sealed queen cells.

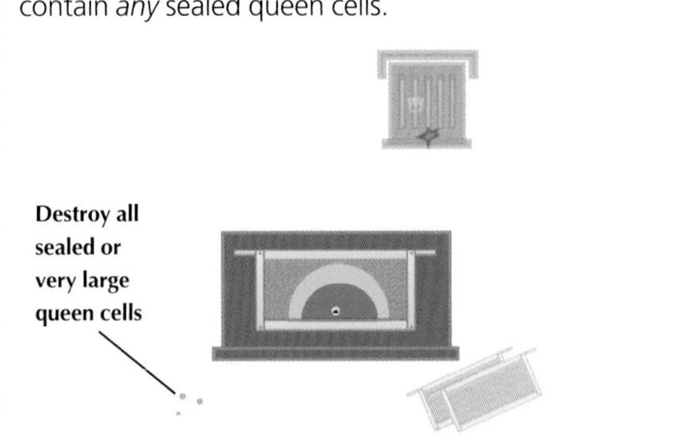

9 Choose one or two which you can see contain a larva and mark the frame(s). One of the easiest ways is to stick a drawing pin into the top bar vertically above the queen cell and towards that side of the frame.

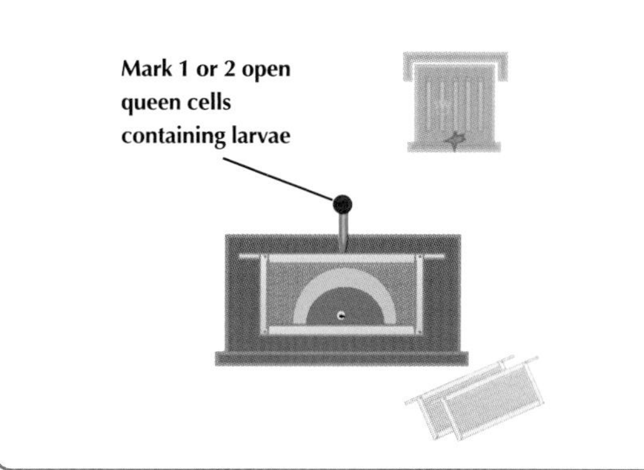

10 Push all the frames together and fill the gap with the frames originally removed from the nucleus box.

11 Replace the queen excluder, supers, inner cover and roof.

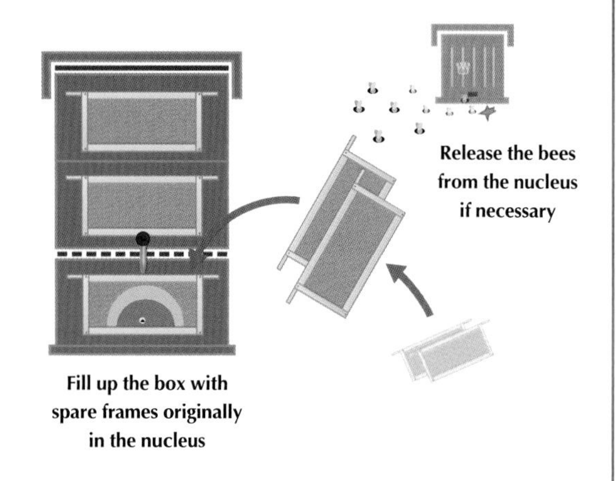

12 If the grass plug in the nucleus has not withered to release the bees in 24 hours, remove it at dusk.

13 NB Because the queen was still present, there will be eggs in worker cells. Bees can rear a queen from a larva that is less than three days old. This means that for the next six days the bees can start rearing more queens on this brood. There's no point in destroying the queen cells that you haven't chosen (see point 9), as the bees will simply build more. If you leave unsealed queen cells in the swarming colony it will be at least eight days before a virgin queen emerges. Even if a queen cell is sealed as soon as you reassemble the hive, the larva still needs eight days to pupate and metamorphose into an adult.

14 Seven days after you've transferred the queen into her nucleus, examine the frames in the swarming colony. When you've removed the outside frame, move the others so that you can inspect those containing the queen cells, which you marked with drawing pins. Don't shake the bees off these combs as this could seriously damage the health of the developing queen! Rather, brush them off with a goose wing feather or a handful of grass and inspect them. Choose one and remove the drawing pin marking the other one(s).

15 Examine all the other frames and destroy *all* other queen cells that you find. You can shake the bees off these frames. (You need to remove them so that you have a good view of both sides of the comb.)

Queen cells built on the face of the comb are obvious, but bees have a nasty habit of building them in nooks and crannies. These can be along the bottom of the comb, squashed in between the comb and the bottom bars or tucked in between the comb and the side bar.

As a beginner it's easy to mistake some drone cells for queen cells. If in doubt, destroy them. It's better to lose a few drones than to overlook a queen cell. If there are other occupied queen cells in the colony when the virgin queen emerges from your chosen queen cell, she may well leave the colony with an afterswarm, as a replacement queen can still be raised from the remaining queen cells. It's therefore essential to make sure that you leave only *one* queen cell.

16 After three weeks, open the colony to check that the new queen has successfully mated and is laying eggs in a good brood pattern. If anything has gone wrong you still have the original queen in the nucleus box and she can be reintroduced to the now queenless colony. To do this, unite her nucleus (and her) to the queenless colony over newspaper.

A queen cell built between the comb and the bottom bar.

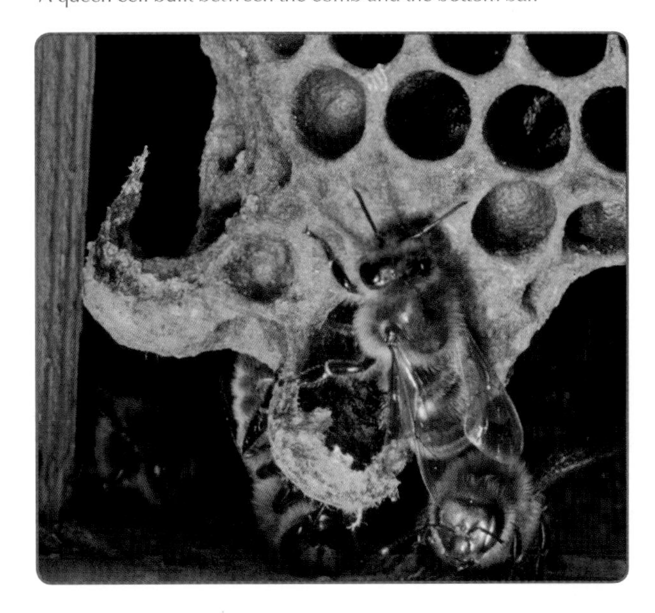

'Artificial swarm' method

This method separates the queen and the flying bees from the brood. You'll need an extra floor, brood box with its full complement of frames with drawn comb or foundation, inner cover and roof. You perform the artificial swarm when you find eggs or larvae in queen cells.

1 Lift the hive containing the swarming colony temporarily to one side. At this point you've separated the brood and the queen from the flying bees, which are operating from the original site.

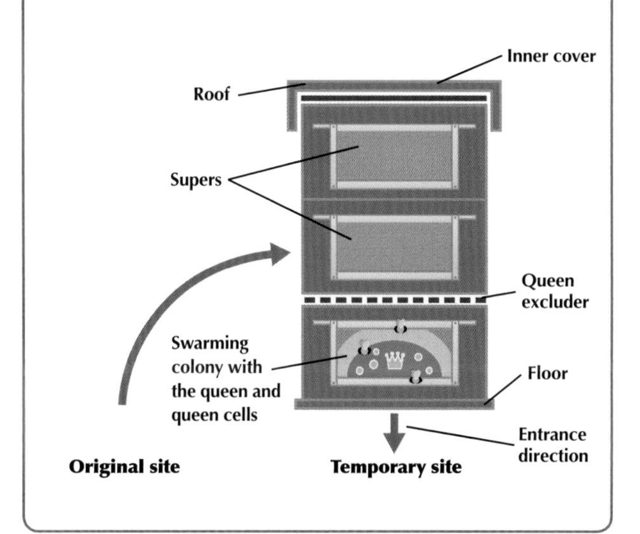

2 Place the spare floor, brood box and frames on the original site.

3 Remove the third and fourth combs from the front (if your frames are warm-way) or from one side (if they're cold-way).

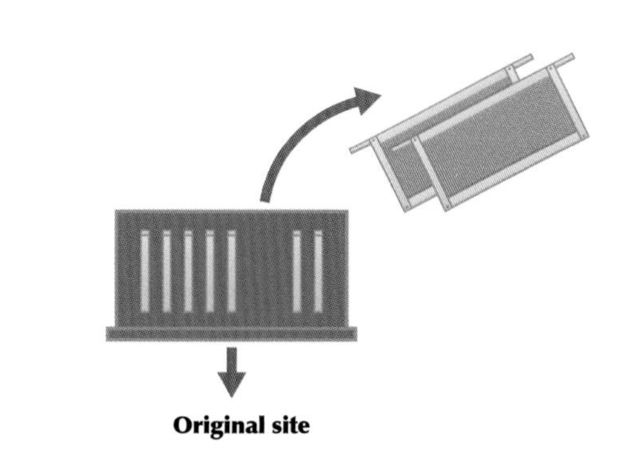

4 Find the queen in the swarming colony. Place the queen, on her frame with the adhering bees but no queen cells, into the new box, which is situated on the original site. Add a frame of brood with *no* queen cells. If any of your spare frames contain drawn comb, put these either side of the queen's frame. If you've no spare drawn frames, take a couple from the flanks of the original box as these are unlikely to contain any brood. However, do a double-check to make sure that no queen cells have been started on these combs. If they have, destroy them. You've now separated the queen and the flying bees, in the new box, from the brood, which remains in the original brood box.

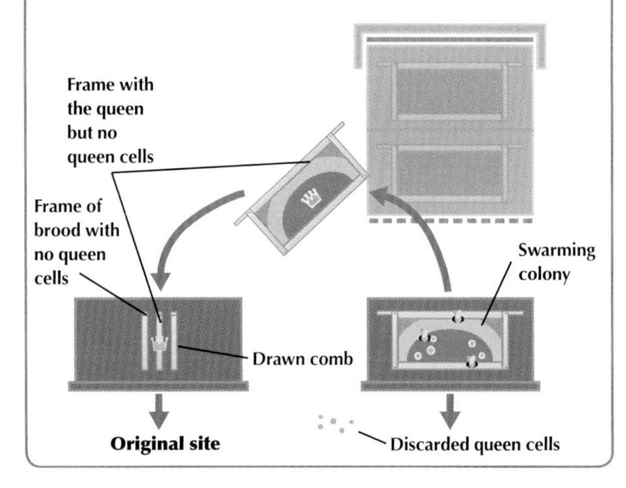

5 Fill up the new brood box with frames. Add the excluder and the supers, then the inner cover and roof.

6 Fill the gap in the original box and position it to one side of the original site with its entrance at 90° to its original orientation.

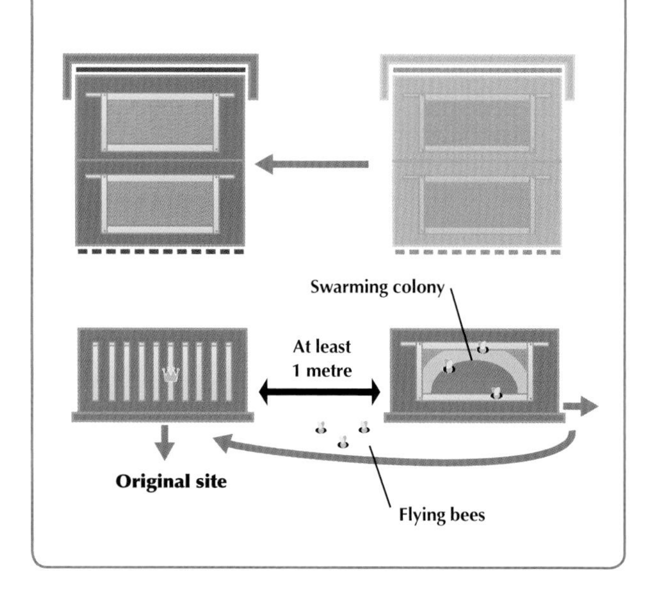

7 Seven days later, move this box to a stand elsewhere in the apiary. This should be to at least one metre on the other side of the original hive, with its entrance facing in a different direction. The bees that have started foraging during the week will have learnt to return to the hive in its new position. They'll now find it gone and will join the flying bees working from the hive on the original site.

8 The colony with the brood, now in its second position, will become depleted of bees. It's queenless so it will be building queen cells, but because of the lack of bees it will not swarm with the first virgin to emerge; she'll kill any other virgins that emerge and the house bees will tear down any remaining queen cells.

You'll need to keep an eye on the colony containing the original queen, because when the colony has built up its strength she may well have another go at swarming.

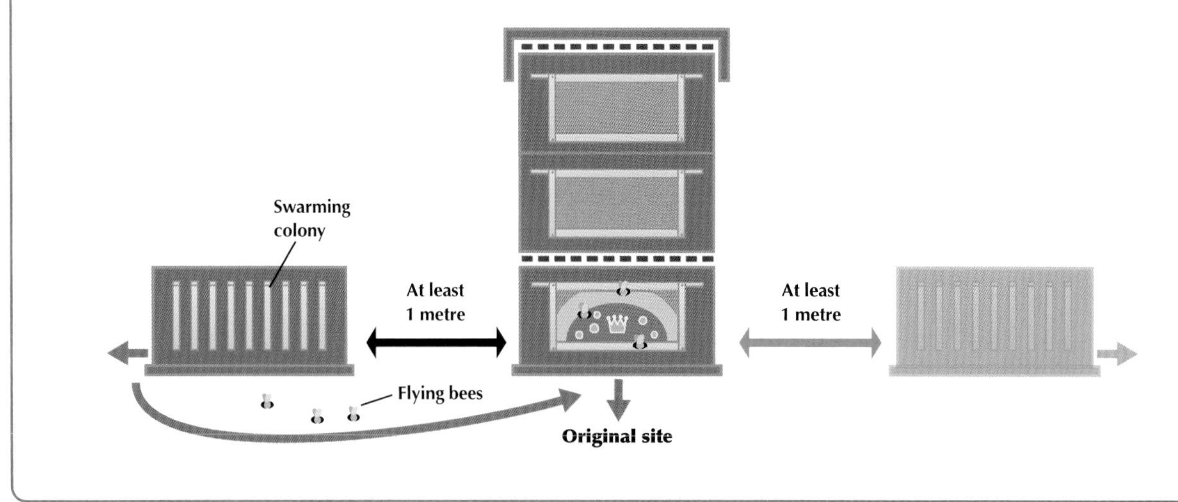

Swarm control without finding the queen

There will be times when you simply can't find the queen. It will be easier if she's marked but even then she may hide. However, all is not lost, as it's possible to control swarming without finding the queen. This is a variation of the 'artificial swarm' and achieves the same rearrangement of the colony.

1 Lift the brood box containing the elusive queen off its floor and place it to one side on an empty super or an upturned roof.

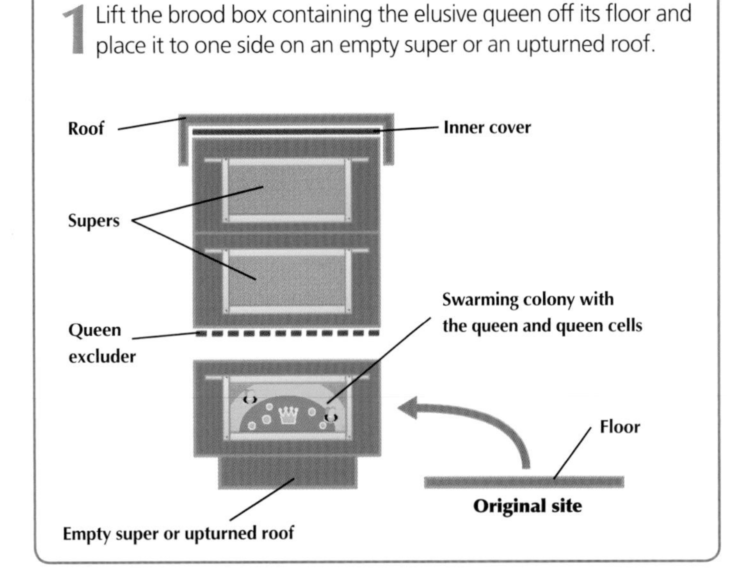

2 Put the new brood box on to the original floor on the original site.

3 Remove six or seven frames from the centre, leaving a large gap. You're going to shake all the bees from the original brood combs into the new box. The queen will be in one of three places – among the bees you shake from the combs, with those on the original floor, or on the side of the old brood box left behind when you removed the frames.

4 Take each comb from the original brood box in turn. Hold it well down inside the new brood box and shake it firmly to dislodge the bees. Brush any bees remaining on the comb into the new box.

5 You do not want bees repopulating the combs you've cleared. Place them in a spare brood box and cover the top with a cloth or inner cover. If you don't have a spare brood box, use a cardboard box as a temporary store, again keeping it covered. You don't need to hang the frames in the cardboard box but it should be large enough to hold all of them. Make sure you keep the frames in the same order as they came out of the original brood box, with the same faces together.

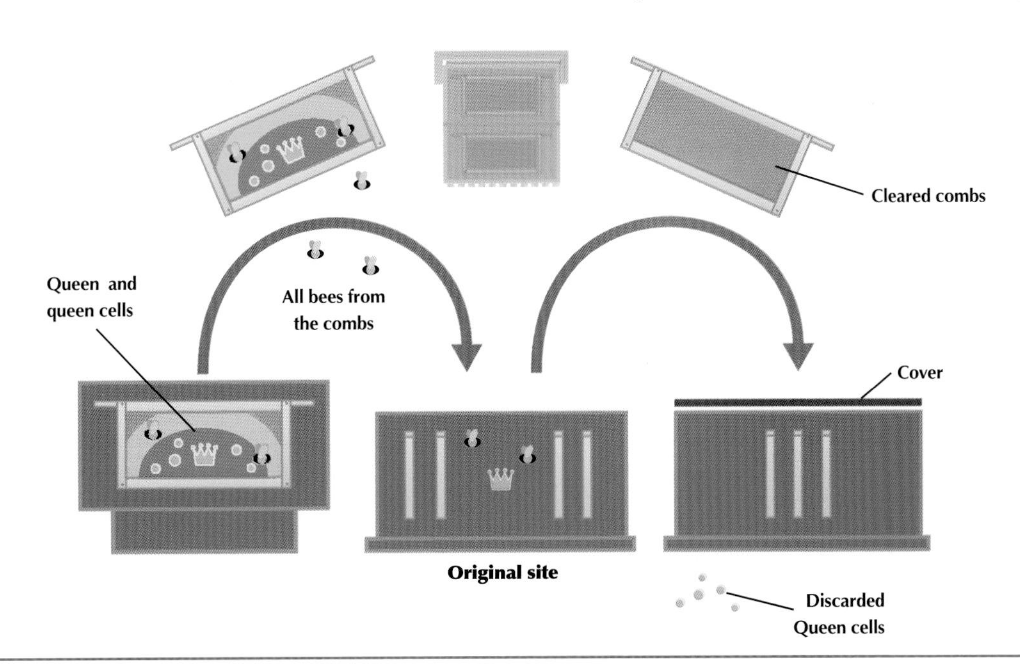

6 As you deal with the combs, look for one with no queen cells but which contains eggs, larvae and sealed brood. Transfer this into the new box rather than the temporary store.

7 Having cleared each comb in this way, take the original brood box, hold it over the new one and brush the bees off into it. The queen should now be in the new box with all the bees.

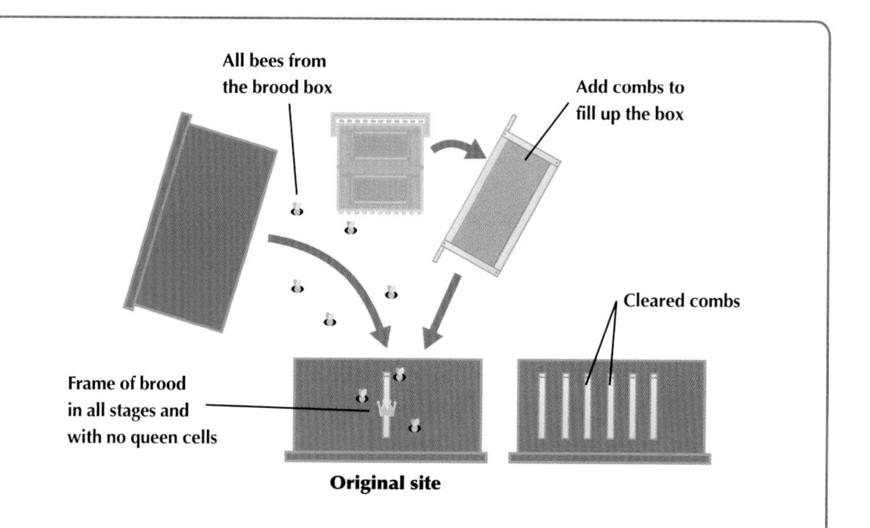

8 Replace the excluder and the supers but not the inner cover.

9 Take the cleared brood combs and replace them in the original box in the right order. Fill up the final gap with the 'spare' frame out of the new box.

10 Place the original box on top of the super(s), followed by the inner cover and roof. Leave your spare 'new' roof, inner cover and floor at the apiary.

11 Twenty-four hours later, nurse bees will have been attracted back to the brood and will be in the top (original) box.

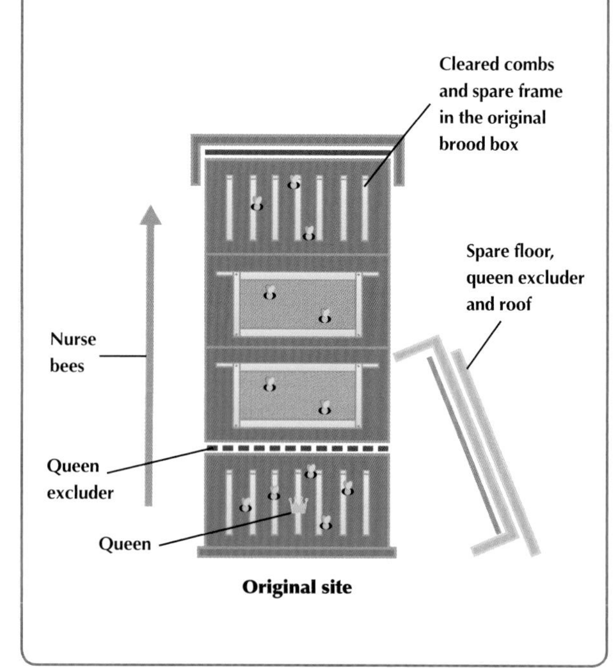

12 Return the following day and deal with the top box. Set up the new floor in its chosen position and place the top box on it. Put the spare inner cover on the original hive and put the roofs on both hives.

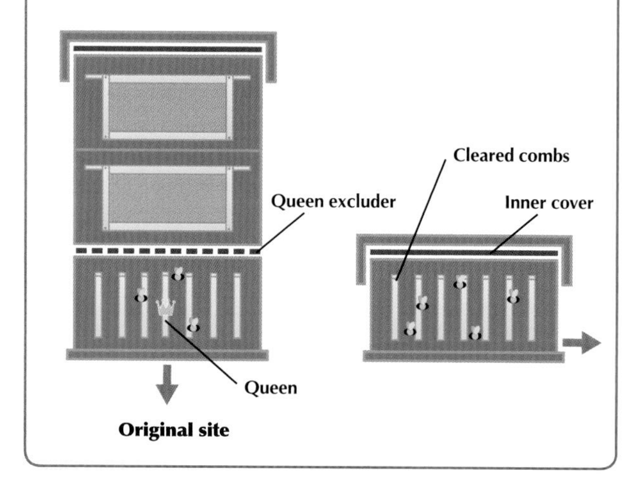

You have now completed steps 1 to 7 of the 'artificial swarm'. The queen and the flying bees are operating from a new brood box on the original site and this hive still has the supers. The brood is in the new box on a different site. You can now complete your swarm control from step 8 of the 'artificial swarm' method.

There are many ways to control swarming bees. You can in fact work out your own method if you take into account the timetable for queen development and remember the principle of separating one out of the queen, brood and flying bees from the other two.

If you find queen cells containing eggs or very small larvae, you do have some time and immediate action isn't required. You have a couple of days to get some help. If nothing else, you have the time to think!

Steps to take if you have lost a swarm

If you inspect your colony and find sealed queen cells, you may have lost a swarm with the old queen. The first thing to do is to look through the colony to see if you can find her. If you do, then proceed from step 5 in the artificial swarm method above.

If there are more than just one or two sealed queen cells, check to see if there are eggs in worker cells. If there are none at all, just larvae, you can be sure that the queen (and the swarm) has left. Another sign that the colony is queenless is that the bees are often more bad-tempered without her.

If the queen has gone and all the queen cells in the colony are sealed and any larvae present are very large, reduce the queen cells to just one large one. Don't let anyone tempt you to leave two 'just in case'. If you do, you are very likely to lose another swarm with the first virgin queen of the two that hatches.

If you find both sealed and unsealed queen cells, remove all the sealed ones and check again one week later to reduce the remaining queen cells to just one. This virgin queen will emerge and then fly to mate, return and head the colony.

Above: If your colony has swarmed you can open a queen cell and release the virgin which is ready to emerge.

Below: An open queen cell from which the virgin has emerged.

If you find one or more queen cells open at the tip with nothing inside, your colony has swarmed and started to produce afterswarms as the virgin queens emerge. You then need to check all the queen cells. You should find some that are 'ripe', *ie* the workers will have removed the wax at the tip and revealed the brown parchment-like cocoon. You may already have a virgin queen loose in the colony.

Take your hive tool or a penknife and use the corner to open the tip of the queen cell. You will find one with a living queen inside ready to emerge and you can release her. This ensures that your colony has a potential new queen. Then you need to remove all the other occupied queen cells. Whatever you do, don't leave sealed queen cells and a virgin queen in the same colony. A colony will produce an afterswarm if it has plenty of bees, a virgin queen loose in the colony and occupied queen cells. If you remove all the queen cells after you release the virgin queen, the colony will not produce an afterswarm, even if another virgin queen is already present. The other factors that will prevent afterswarms are if the colony has few flying bees or if it is weak.

Two weeks later, check your colony to make sure that your virgin queen has mated and started to lay. The colony is then back to normal.

Colony uniting and increase

When we were considering swarm control earlier in the chapter, we noted that swarming is reproduction at the colony level. The outcome of the artificial swarm is two colonies of bees, one with the original queen and one with a new queen. If you kept both colonies, repeated the process for each of them every year and none of them ever died, this simple doubling process would rapidly make you the largest beekeeper in the world!

However, colonies of bees do die for a variety of reasons. Of the ones that survive, by no means all will be colonies that you want to keep. A good time to consolidate numbers is after the end of the summer honey season, when you've removed the supers, and before any autumn honey flows, such as from ivy or heather.

Consider the artificial swarm. You end up with a colony headed by the mother (original) queen and one by her daughters. Perhaps you only have room to keep one of them so choices must be made. You know how the mother's colony works and behaves. Queens lay less well as they get older but some queens can head productive colonies into their third year, so you may want to keep her. On the other hand, the daughter's colony may show better potential. It may be better-tempered, for example. If you're able to manage both colonies side by side for a while you can make a comparison.

My advice is to kill the queen you don't want to keep. This may be obvious but your records or a gut feeling will tell you which one you would prefer to head a colony in your apiary next year. Kill the unwanted queen before you begin the uniting process.

Below: Killing the queen.

The most humane and rapid way to kill a queen is to crush her thorax firmly between thumb and finger. If you feel unable to do this, put her into a container and place this in your freezer for 24 hours. She will become immobile with the cold and die quietly.

Uniting

You now have two colonies – one that's described as 'queenright' (*ie* with a laying queen) and one that's queenless – which need to be united into one. This will necessitate moving at least one of the hives and possibly both of them.

When a hive is moved, the returning foragers go back to its previous location. After a while they begin to look for their colony. If they don't find it, they approach the nearest hive and behave submissively to the resident guard bees asking to be let in. The newly arrived bees expose their Nasonov glands and fan the pheromone into the air, attracting their nest mates to the hive, and shortly things settle back to normal. However, you'll probably find that a few bees will continue to check the old site for many days before accepting the new hive as home.

If there has been some robbing going on the guards of the nearby hive may be very defensive. They may not accept the submissive posture of the new bees, and some of these may be killed. You can help to prevent this by moving the receiving hive into a position midway between the two sites. This effectively 'confuses' both sets of returning bees at the entrance and definitely reduces fighting there.

Uniting over newspaper

This involves placing one hive on top of the other with a sheet of newspaper in between, and is one of the most effective uniting methods. The newspaper must be large enough to cover the top of your brood box completely. If a single sheet isn't large enough, join two together with masking tape.

Open the hive that will be at the bottom during uniting and scrape the top bars of the frames in the brood box clean of wax and propolis. Spread the newspaper over the top, fold over the corners and secure the paper to the sides of the box with drawing pins. You can also hold the paper in place with a queen excluder. Otherwise, even on a very calm day, you can be sure that just as you let go there'll be a gust of wind to blow the paper away!

Above: Placing newspaper over the lower brood box.

Although not strictly necessary, many beekeepers make a few very small slits or holes in the newspaper using the corner of their hive tool. This is believed to encourage the bees to start chewing through the paper.

Above: Making small slits in the paper.

Above: The wind often blows it away.

In the hive that's to go on top, clean the bottom bars of the frames to remove any brace comb or other projections that might tear the newspaper. Then place this hive carefully on top of the newspaper. If the boxes aren't aligned properly with the outside surfaces flush, don't try to slide one on the other – this will tear the paper, which can lead to fighting between workers from the two colonies as they make contact too quickly. Colonies will unite peacefully when they're given time to get used to the odour of the other colony as they slowly chew away the paper barrier.

Above: Cleaning brace comb from the bottom bars.

Above: It can be held in place with a queen excluder.

Above: Adding the top brood box.

Above: Shaking in the remaining bees.

Above: Adding the inner cover.

Above: Adding the roof.

Right: Bees chewing through the paper.

Leave the colonies in position for one week. Then you can go through the boxes, lifting the top one off first and placing it on the upturned roof. You need to rearrange the combs. Put all the combs containing brood in the bottom box and all the others in the top one. Food combs should be positioned over the brood nest in the bottom box. The bees can overwinter like this and you can reduce the colony to a single brood box in the spring. Often there will be enough food (honey) from the two boxes to provide most of the winter feed required by the combined colony during the winter. There should certainly be a good proportion of the necessary winter stores. With prolific bees, you may find that there are too many brood combs for one box. If so, leave things as they are. Such a colony may well be more suited to a double brood-box system of management.

General rules for uniting

- When uniting a queenright colony with a queenless one, the queenless colony goes on top.
- When uniting colonies of different sizes, such as uniting a nucleus to a full colony to re-queen it, the smaller colony (the nucleus) goes on top.
- When you move one colony to unite it to another staying on its original site, the moved colony goes on top.

To unite a nucleus to a full colony, first transfer the frames from the nucleus into a full brood box. Fill the gaps with frames of foundation or drawn comb.

Uniting summary

1. Kill the queen you don't want.
2. Clean off the top bars in the bottom box and the bottom bars in the upper box.
3. Cover the whole of the brood box area below with newspaper.
4. Place the queenless, smaller or moved colony on top.
5. Replace the inner cover and roof.
6. Rearrange the frames one week later.

Colony increase

If you want to increase your number of colonies and you're prepared to feed your bees and take no honey crop, you can easily expand from one colony to five or six. However, this needs skill and experience so I would recommend that you start on a smaller scale.

You can do this easily by using the artificial swarm to control a swarming colony. This involves leaving the queen and the flying bees on the original site and moving the brood nest to one side. A week later, this box is moved away. If you let these bees rear a queen and she mates successfully, this will give you two colonies. If you want more than two, instead of moving the box of brood a second time you can divide it, I suggest into no more than three parts. The number of divisions will partly depend on the amount of equipment you have available. Each division needs a home, which can be either a nucleus hive or a whole brood box with its associated floor, inner cover and roof.

Dividing a colony after an artificial swarm procedure

1 Set up the required number of boxes near the box of brood that is to be divided ('Box A'). This is the brood from the swarming colony, which has been moved to one side from its original site.

2 Check the brood combs for queen cells. You'll need a frame of brood with at least one good queen cell for each division.

Below: Select a good sealed queen cell for each division.

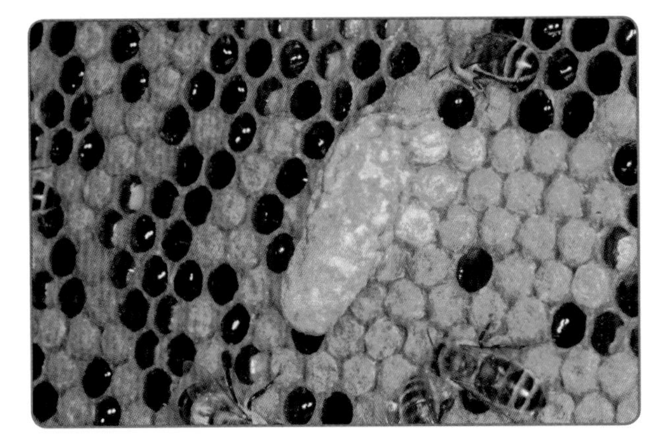

3 Put one comb of brood and its adhering bees into each division, leaving one in Box A.

4 Put at least one comb of food into each division. You don't have to leave a comb of food in Box A as this retains the flying bees, which will continue foraging. You'll have too many brood and not enough food combs when making up the divisions.

Below: A frame of food.

5 If there are insufficient food combs in Box A to give one to each division, you can use a food comb(s) from your other colonies. Take a brood comb from Box A and shake the bees off into one of the divisions. Remove all the queen cells from this comb. Take a food comb from the donor colony and shake off the bees into their own hive. Replace it with the comb of brood. Place the food comb into the appropriate division. If you do not have another colony that can donate food combs, reduce the number of divisions to the number of food combs available.

Below: Destroy all unwanted queen cells.

6 Box A now contains a brood comb with one selected queen cell and the remaining brood combs with their adhering bees. You now need to distribute these bees between all the divisions, not forgetting Box A. Roughly divide the number of remaining combs by the number of divisions and then shake the bees off the appropriate number of combs into each of the new boxes, returning the cleared frames to Box A. Because Box A retains the flying bees, err on the side of favouring the new divisions when you're distributing the bees. You're aiming to have all divisions containing roughly equal numbers. Because the divisions lose all their flying bees back to Box A, you don't need to destroy any of the queen cells on the brood combs you've given them. There will be no afterswarms.

7 Arrange the combs in the division boxes so that the brood comb is next to the side (or front) wall, followed by the food comb. Fill the box with spare frames.

There are many other ways of achieving increase. To make lots of new colonies you'll need the appropriate equipment, but if you're a new beekeeper this will be at a premium. Remember, each successful division you make will eventually need a floor, brood box, three or four supers, a queen excluder, an inner cover, a feeder and a roof.

Queen Introduction

The whole character of a colony of bees comes from its queen. Change the queen and the whole colony will change. Beekeepers have produced what might be called a 'sub-industry' because of the ability to rear new queens by the thousand and sell them to other beekeepers. Such queens in this country are transported in little cages. Several workers are also included to feed her from a small food supply in the end of the cage. The method? Ordinary letter post! A large proportion of these thousands are killed by the bees to whom the queen is given.

Whole books have been written devoted solely to the introduction of queen bees to a new colony. Some beekeepers regularly requeen their colonies citing 'bad temper' as a reason. The more exotic and different from the normal bees of your area the new queen is, the more likely is this descent into strong defensiveness (aggressiveness from our point of view on the receiving end). This is most likely to happen with strains or races of bees of hybrid origin. These will also deteriorate over the generations as new queens raised mate freely with the local drones. Repeat your mistakes and you will repeat history.

Yet, changing queens is a vital skill. A strain of bees cannot be maintained without culling the queens producing bad colonies and replacing them with those reared from good colonies. So what is the problem? It is that most colonies only have one queen and will tolerate no other. Another is that if the queen dies within a few hours of introduction, bees will start to rear a replacement queen from newly hatched larvae. A beekeeper finding a colony with no brood in August may assume there is no queen. This may not be so. Broodless colonies are not necessarily queenless colonies.

The first thing to do is to make sure that the colony is actually queenless rather than the queen has stopped laying for some reason. Insert a comb from another colony that contains eggs and check seven days later. A queenless colony will have started to raise queen cells on this brood. If there are none, the colony has a queen.

For the best results when introducing a new queen, the colony must be recently queenless, *ie* you have killed the unwanted queen and then introduce another. Colonies that have been queenless for a while are more reluctant to accept a new queen. Any queenless colony will accept a new one more readily if she is in the same state as the one they have lost. In other words, if you kill a queen that is in full lay, the colony is more likely to take to the new one if she is also in full lay. The safest way to introduce the new queen is in a nucleus which you then unite to the dequeened/queenless colony using paper (see page 92).

It may be that you have decided to buy a queen from a supplier. She will come through the post in a cage, accompanied by a few workers, probably in an envelope. When the package arrives, give the bees a few drops of water directly on the mesh of the cage and put it into a warm, dark, quiet place (such as the airing cupboard). Then you could kill the queen that you do not want.

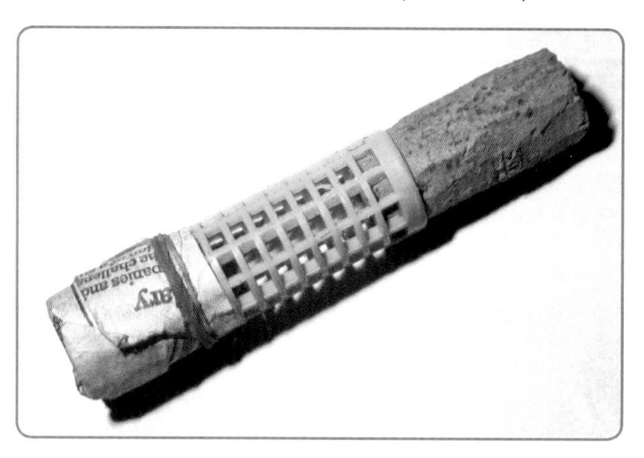

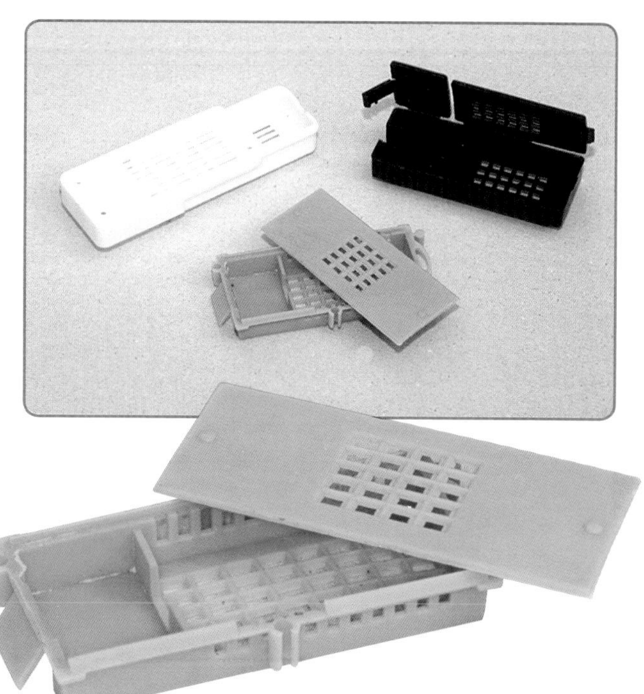

Above: A 'Butler' type cage made from a hair curler.

Above right: Commercial queen travelling and introduction cages come in a number of different designs but all have the same basic structure.

Right: A commercial queen travelling and introduction cage showing the compartments for the queen and attendant workers and the candy.

Above: The queen introduction cage in place.

However, the queen that has travelled in the post will not have been laying eggs for a while and a colony which has just lost its queen will be expecting a queen in full lay. The safest thing to do is to create a nucleus – without a queen – as set out in the nucleus method of swarm control but without the queen (see page 89). Then introduce your queen to that nucleus. A three-frame nucleus in a five-frame box allows frames to be moved apart temporarily to accommodate the queen cage.

Bees try to kill a strange queen by forming a ball of bees around her. They will do this to the queen cage. The safest methods of queen introduction therefore involve keeping the queen in a cage during this period of initial aggression. It is quite possible to use the travelling cage or other similar cages which are available from suppliers. If you have reared the queen yourself, she can go on her own in a mesh tube called a Butler cage. Those of you with hair can use a suitably sized hair curler with newspaper held over each end with rubber bands!

The travelling cage will have the hole at one end plugged with candy to provide food for the bees during transport. The bees will eat this and release the queen in due course. The bees will chew through the newspaper over the ends of the Butler cage and release the queen that way. The aim is to delay the queen's release until the nucleus is quiet and undisturbed. With the Butler cage, I always use enough paper to give the queen a covered area as long as her body. This gives her somewhere to get away from the attentions of workers in the colony if necessary. Introduced queens have been found to be missing feet, presumably pulled off in the initial aggressive period. However, workers in the colony must be allowed to contact the queen in the cage in order to pick up her pheromones so they must have access to the mesh in the rest of the cage.

Check the queen is out of the cage in 24–48 hours and then remove it. Close up the frames but do nothing else. Leave the whole thing alone for one week before examining it to make sure the queen is present and alive.

This is one occasion where you want any flying bees to leave the nucleus so don't block the entrance or do anything to prevent them leaving. The older the bees, the more aggressive they are likely to be to the new queen. On the other hand, young, newly emerged bees will accept any queen.

When the new queen is laying well in the nucleus, kill the queen you do not want and unite the nucleus to your now queenless colony. Many beekeepers will scoff at what seems an over-cautious procedure but success is worth the extra effort. It is even better to wait until the new queen's workers are emerging before uniting the colonies. The more of her own bees the queen has, the safer she will be. If you want to know more, read one of the books devoted to the subject.

To sum up

- Only introduce a queen to a colony you know to be queenless – because you have removed her.
- The queen is safest when first introduced to a nucleus with few or no flying bees.

It is probably true to say that anyone who tells you queen introduction is easy hasn't done it often enough.

Forage

When used by beekeepers, the term 'forage' describes the food sources available to bees. As a verb, it describes the activity of bees working in flowers. Bees visit flowers for their food, consisting of nectar and pollen. They also collect nectar from extra-floral nectaries (nectar-producing points) that a few plants develop separately from those within their flowers. An example is the points on the leaves of laurel that are visited by bees. Bees also collect honeydew, resin from tree buds (propolis) and, of course, water.

Above: A honey bee working an extra-floral nectary on a laurel leaf.

Nectar

Nectar is a sugar-rich fluid secreted mostly in specialised areas within flowers known as nectaries. Nectar produced in one flower species is different from that produced in another, resulting in the different flavoured and coloured honeys that you see on shop shelves.

Above: Nectar in the base of a cherry flower.

Different flowers have developed different structures that ensure that they're pollinated when a bee, another insect, a bird or a bat collects nectar. Mostly the nectaries are deep in the flower and pollen is brushed off on to the pollinator's body as it passes the anthers. When the pollinator visits another flower of the same species, the pollen is transferred to the stigma, effecting pollination and seed production.

Some flowers have nectar guides on their petals. Some of these are obvious stripes, such as on a crocus flower, but others reflect ultraviolet light and are invisible to our eyes. This is because humans and bees see different parts of the spectrum. Unlike us, bees cannot see red but they can see ultraviolet, and thus the 'invisible' nectar guides indicating the position of a flower's nectaries. Nectar guides may also take the form of scent markings that are obvious to the bee, which smells them through its antennae. Flower scents are also picked up by the bee's waxy outer body covering. When the forager returns to the hive and communicates the nectar source by dancing, recruits will be aware of these scents which help guide them to the correct flowers.

In the early part of the year the incoming forage is used to build up the colony population until the foraging bees collect far more on a 'good' day than the colony can eat. This is then stored in the comb. Bees are quite capable of collecting more than 5kg (11lb) of honey in a day.

Honey flow

Lots of nectar-producing flowers blooming in warm weather of 21°C and over, with moisture in the soil, can generate a honey flow. Flying bees that find the nectar source return to the hive and communicate its location to more workers, which join the foragers. The resulting honey crop can be very large. In the 1950s, one Australian beekeeper averaged 700lb (320kg) each from 400 colonies, and I know a beekeeper in the UK who collected two tons of honey from 17 colonies. However, such results are unusual. In Britain the average honey crop per hive per year is 20–30lb (9–14kg).

The bee dance

Scout bees, which make up about 5% of a colony, go out looking for new sources of food. They then have to communicate its location to the remaining foragers in the hive. They do this with a combination of scent, sound and movement, and this dance 'language' of bees was first decoded by Professor Karl von Frisch. There are two main forms: the round dance and the waggle dance.

The round dance

In this dance the scout bee runs round in a circle and, on completing the circuit, turns and runs back the other way. She repeats this many times and attracts recruits. Her waxy exterior has picked up scents from the flowers and these are learnt by the recruits so that they know what they're looking for. The scout is simply saying 'Go out of the hive and look around within a radius of 10m (30ft).'

When bees are taking syrup from a feeder they perform the round dance, which is why other bees readily find any syrup you may have spilt on the hive or the ground. Syrup has a scent and a careless spill can encourage robbing.

Variations of the round dance indicate food sources up to 100m (300ft) from the hive.

Above: A worker performing the waggle dance.

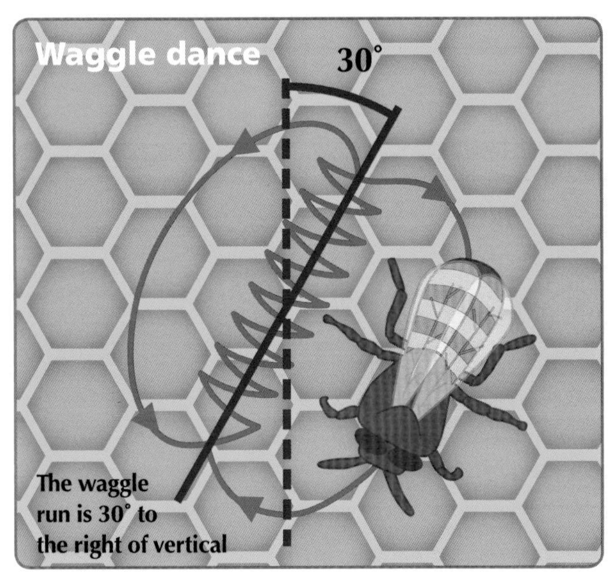

The waggle dance

This is more complicated and communicates the direction, distance and nature of food sources that are over 100m (300ft) away. The scout bee stands on the comb and waggles her abdomen at about 15 'beats' per second. She then runs in a straight line, turns and runs back to the starting point in a rough semi-circle. She waggles her abdomen again before running along the same straight line. However, this time she turns the other way to return to the beginning. As she repeats these actions, interested bees will touch and follow her. They get information about the type of food from the scent on the dancer's body. The vigour of the waggle tells them how rich the food source is and the number of runs per 15 seconds indicates how far it is from the hive, with a lower number indicating a greater distance. The recruits make almost inaudible buzzes. When the scout bee hears these, she stops and feeds them some of the nectar she has found so that they know what they're looking for.

The straight line run indicates the direction of the food source relative to the hive entrance. The dance takes place in the darkness of the hive, so the bees use the sun as a point of reference before they fly off. In the hive, the 'direction' of the sun is vertically upwards. If the scout has found a source 30° to the right of the sun, she'll run at a line 30° to the right of the vertical. The really amazing aspect is that as the sun moves across the sky, the dancing bee alters the angle of her run.

The bees' honey requirements

A colony needs something like 136kg (300lb) of honey per year just to survive from day to day. The beekeeper only sees the surplus of 14kg (30lb), but you need to remember that this is on top of the basic rations. This certainly puts the occasional surplus of 90–140kg (200–300lb) in perspective. As far as I know, the 'official' British record surplus honey crop is 199kg (439lb) – truly phenomenal!

Honey sources

So, which plants produce our honey? Lists of the plants worked by bees are surprisingly consistent throughout the British Isles. All the plants have a flowering period and the same plant will begin to flower early in the south of England and progressively later as you travel north. The same effect occurs with increasing height above sea level, with plants at the bottom of the mountain flowering before those higher up.

Some areas in the British Isles have local conditions and flora which produce a unique, identifiable honey. One of the commonest specialist honeys is ling heather, from *Calluna vulgaris*, or bell heather, from *Erica cinerea*. Other specialist crops include borage (*Borago officinalis*) and sea lavender (*Limonium vulgare*).

Below: Heather.

Below: Borage.

Oilseed rape

Brassica napus is an agricultural crop cultivated in the UK since medieval times. It's grown for its oil, which is crushed from the seeds. Forty years ago it yielded nectar well, but as far as the beekeeper is concerned it's less of a major honey source than it was, although it can still be a major player in warm springs. It's grown countrywide but especially in arable farming districts.

Above: Oilseed rape.

In the 1960s honey could be left on the hive until the end of July and it would remain liquid and easy to extract. Honey containing a high proportion of dextrose granulates (crystallises) very easily, and oilseed rape honey is a prime example; it will also affect most other honeys that contain some rape nectar. It can granulate even before it's extracted, and solid combs have to be destroyed to remove the honey. The modern practice is to extract early honey as soon as enough of it's capped over and hopefully before it sets. This means extracting between mid-May and mid-June. Just when you think you've successfully extracted your oilseed rape honey, you may be thrown if the bees discover a field of spring-sown oilseed rape that flowers in June!

At the same time as the oilseed rape, bees could well be working members of the Acer family such as Norway maple (*Acer platanoides*), field maple (*Acer campestre*) and sycamore (*Acer pseudoplatanus*). Fruit trees are also flowering at this time together with currants – both black (*Ribes nigrum*) and red (*Ribes rubrum*) – and gooseberries (*Ribes uva-crispa*).

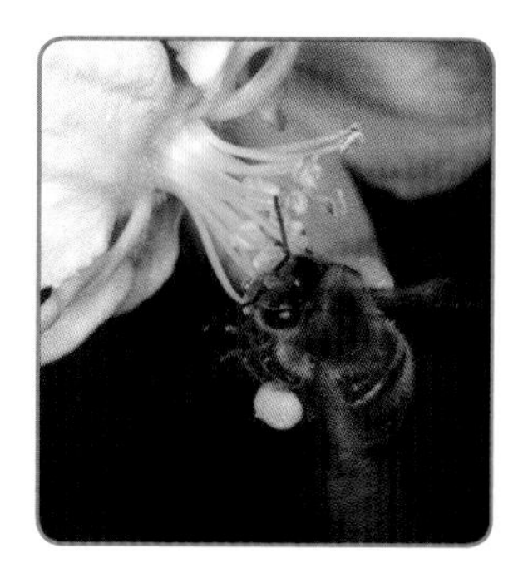

Above: Apple.

Bees work a range of flowers and it's the combination of nectars from different species in various quantities that gives honey its variety of flavours.

The June gap

Although it's less evident in these days of global warming, many districts have a gap between the mass of floral sources in the spring and those of high summer. This is known as the 'June gap'. In your area it may be very definite or you may not notice it at all.

In urban districts there's always something for bees to work at any time during the active season from April to September, and older town suburbs can be excellent foraging areas.

The main flow

Late June marks the start of the 'main flow'. Trees are still good sources, with types such as lime (*Tilia cordata*) and sweet chestnut (*Castanea sativa*) both yielding nectar well when the weather conditions are right. In some areas summer flows end in the middle of July. In others, blackberries (*Rubus fruticosus*) and flowers such as rosebay willowherb (*Chamerion angustifolium*) can continue yielding into August. As the honey flows dry up bees become more inclined to try to rob one another's colonies. They can also become more touchy. Fine weather helps, but in spite of late flowers there may be little return for the bees' efforts.

Above: Sweet chestnut.

Above: Blackberry.

Heather honey

Ling heather (*Calluna vulgaris*) starts to flower at the end of July in the south to mid-August in the north. Some beekeepers take advantage of this late crop by moving their bees to the heather moors. Strong colonies are strapped up with a travelling screen to give ventilation and taken to the crop. I suggest that you gain some experience before you venture into migratory beekeeping. It involves different techniques and equipment, although it can be really enjoyable. Because heather honey is thixotropic (*ie* less viscous) it's often sold as 'cut comb'.

Above: Hives on the heather.

Ivy

The last nectar source of the year is ivy (*Hedera helix*). This starts to flower around the end of September and in warm autumns can yield well. If you've fed your bees for the winter you may well have to give your colonies a super to provide room for them to store the crop. Ivy honey granulates quickly and sets hard like oilseed rape, so if you want to extract it you need to check when it's ripe and deal with it then.

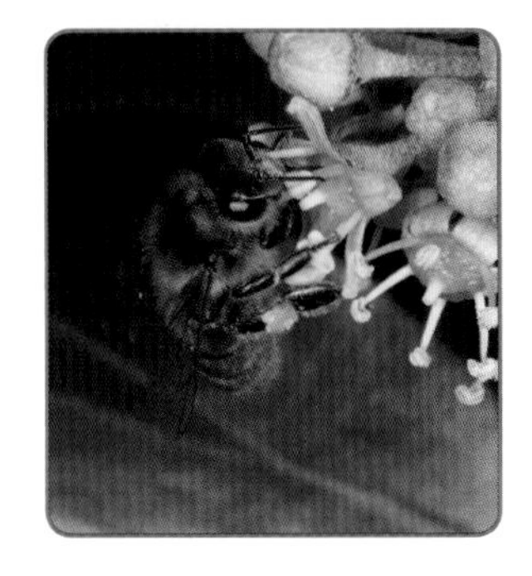

Above: Ivy.

Honeydew

Bees also collect honeydew, a sugary liquid expelled by aphids and other sap-feeding bugs such as scale insects. The droplets land on leaves and other surfaces and are diluted early in the morning by dew. Such 'flows' are usually over by 8:00am and resume the following morning.

Bees will collect honeydew from trees with heavy aphid infestations, such as plums, oak and beech. The main honeydew producers in the pine forests of Germany are scale insects. Honeydew can be very dark with a very pleasant taste.

Bees are very interested in most sugar solutions, but they must contain at least 10% sugar. Below this bees aren't interested, even if they're starving. Most nectar seems to be in the 30–40% concentration range. At low concentrations bees can get very excited when they find a nectar source but this frenzied response diminishes as the sugar concentration increases. This applies equally to a sugar solution given to the colony in a feeder.

The influence of the weather

However many flowers there are, the weather is the final arbiter of what happens. A honey flow needs flowers combined with warm weather and a good level of moisture in the ground. Sun is not absolutely essential. Indeed, the best honey flows can occur on dull, overcast, humid days.

The observant beekeeper will be the first to recognise a honey flow. Activity at the hive entrance greatly increases. Returning bees fly heavily, their legs held forward to counterbalance the weight of nectar in their abdomens. They'll hurry into the hive. There is an air of activity and bustle about the whole colony. Supers will be filled rapidly, so check if more are needed.

Pollen

In a year, a colony needs to collect and eat about 45kg (100lb) of pollen. This is high in nitrogen and is the bees' source of protein. However, no single pollen delivers the full range of protein requirements, so bees with access to only one type of pollen would suffer from malnutrition, just as we would.

If you look after them properly your bees will have adequate food reserves in the spring, but they do need fresh pollen. Many of our native early flowers invest a lot of effort in pollen production but their nectar production is often poor. Indeed, some flowers like the field poppy seem to produce no nectar at all.

Bees will forage primarily for pollen when the colony needs it. You'll notice different coloured pollen loads being taken into the hive and stored in the comb. Books that show the colours of pollen from different flowers will enable you to identify the flowers your bees are working.

Below: Pollen colours indicate which flowers your bees are visiting.

Supering

To a beekeeper, the term 'supering' means providing additional room as the bees require it. The queen is confined to the brood box by a queen excluder, above which space is supplied by the addition of extra boxes called supers. Because the queen cannot enter the supers, the comb in them is free of brood and honey is stored in cells that have not been used for brood rearing.

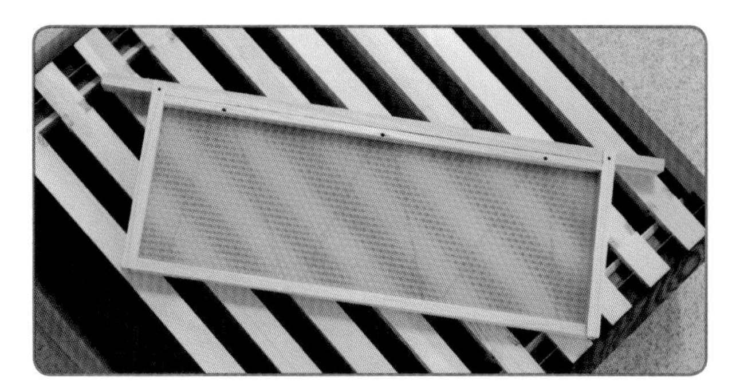

Above: A super frame with wired foundation.

There is no law that says brood and super frames cannot be the same size. However, most beekeepers in the UK use shallow boxes and frames for their supers. The frames are known as SN1 and are as wide as the DN1 brood frames but only 14cm (5.5in) deep. I think they're easier to handle and uncap to remove the honey, but you need more of them than if you were using brood-sized frames for honey storage. For most seasons you'll need three or four supers for each colony that overwinters.

A lot of early honey collected by bees today tends to granulate in the comb if it's left on the hive too long. Shallow supers give more flexibility in removing part of the crop when it's ready but before it granulates. Another consideration is weight. A deep National frame holds about 11kg (5lb) of honey. This means a very full brood box used as a super will weigh 121kg (55lb), plus the weight of the frames and the box. A full shallow super holding 55kg (25lb) of honey plus the frames and box is quite enough for most of us. Start with shallow supers and see how you get on.

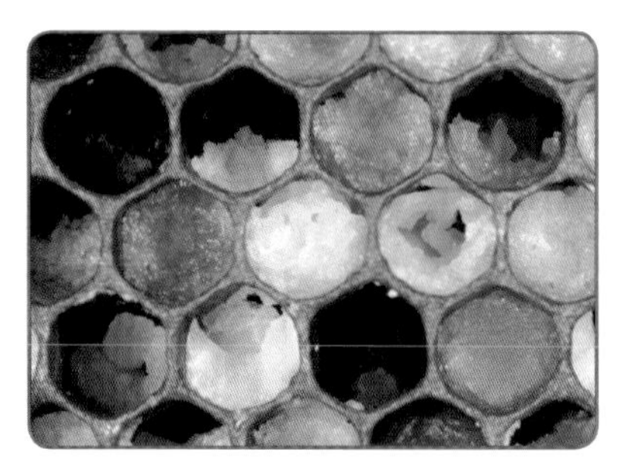

Left: Honey granulated in the comb.

Bees store honey differently throughout the year. In spring, when the brood nest is developing, they want to store it away from the brood area. A colony will want more room during this time than you would think. So when should you put the first super on your hive? The simple answer to this is 'When the brood box is full of *bees*.' Likewise, when the first super is full of *bees* add a second.

If you've been feeding your bees heavily in the spring to stimulate colony development, they will have stored some of this sugar. The first super can fill quite quickly with stores that the bees transfer from the brood box. There's therefore a chance that you'll have a mixture of sugar and honey rather than pure honey in your first super. The answer to this potential problem is to feed in the spring only if the bees really need it.

The first super

If you decide your bees need a super you'll also need a queen excluder, which should be clean of brace comb and propolis. When you scrape a queen excluder, make sure it's lying on a flat surface like a hive roof. Take care and don't catch the slots. If you deform them they may be just large enough for the queen to get through.

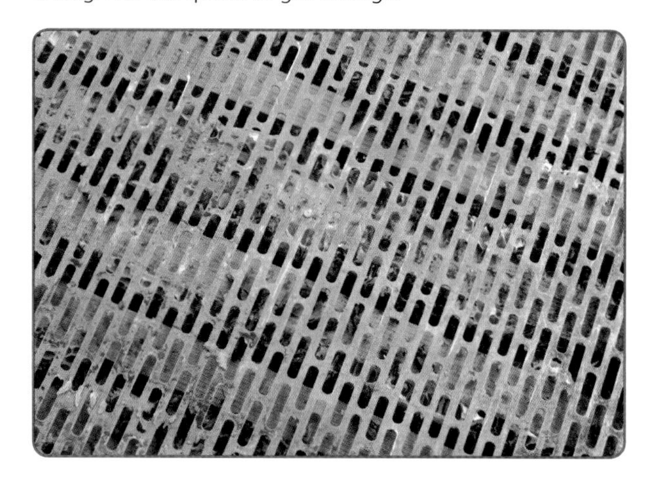

Above: A zinc queen excluder.

Open the hive in the normal way and make your examination. Put the combs back in order and push them together. Clean the wax and propolis off the top bars and put on the excluder, followed by the first super, the inner cover and roof.

Above: The excluder in place on a colony on brood-and-a-half.

Bees can be a little reluctant to draw out foundation at first, but when they need new comb they'll move into the super. In due course they can be encouraged to move up into the second one by taking a couple of drawn combs from the first super and swapping them with frames of foundation from the second, putting the drawn combs over the greatest concentration of bees in the first super. Both supers should still have a full complement of frames.

In a good season, the bees will need a third super. When they've started to occupy it you can inspect the others. Some honey can probably be removed, maybe the whole of the first super. However, reducing the amount of space the colony has can cause congestion, which is where your fourth super comes into play. Put the first super to one side and return the second and third to the hive. Put the fourth on top followed by a 'clearer board' with Porter bee escapes in place and then the first (full) super. Complete the hive with the inner cover and roof. The bees still have access to three supers and have not lost any room. Methods for clearing bees from supers will be covered in Chapter 4.

Left: Porter bee escapes in the clearer board.

In areas with a good amount of forage, given the right weather conditions, bees can fill a super of drawn combs in less than a week. If times are good, don't be afraid of putting more than one super on at a time. If they don't need the second one the bees will ignore it, but if it isn't there they cannot fill the combs with honey.

'Drawn' supers, that is supers containing frames with drawn comb, are filled very quickly. It's only in your first year of beekeeping that you'll be in a position where you have to give so much foundation to your bees.

By late June/early July bees may be storing honey well, but a change is beginning to take place. The queen's egg-laying rate is decreasing, the size of the brood nest is diminishing and the urge, with our native bees at least, is to begin to consolidate. By this time of year you should make sure that supers are well filled before you add another. As the brood nest reduces the bees may start to store honey in the brood box, which means you'll have to feed less later on. Yellow Italian bees are far less inclined to do this than dark native bees and they'll need more food to survive the winter. Indeed, once the supers have been removed such colonies may have no food at all and require immediate help in the form of sugar syrup.

If your bees are near heather (*Calluna vulgaris*) moors, remove the super(s) containing the summer honey and give the colony a fresh one for the heather honey. This has a strong, distinctive flavour and is thixotropic, with a jelly-like consistency. As was mentioned earlier, it's often sold as 'cut comb', where pieces of the whole comb, including the wax, are eaten. Obviously, if you plan to produce cut comb you need to fit your super frames with unwired foundation! You can then cut the combs into pieces and sell them in appropriate containers.

Right: Heather honey in the jar.

Right: Unwired foundation.

CHAPTER 4 BEE MANUAL

THE HONEY HARVEST

The honey harvest

Bees collect honey for use by the colony as winter stores, enabling it to survive until the following spring. Beekeepers care for their bees and, in return, take a share of this harvest. In some years many bee colonies would die if they weren't looked after by beekeepers.

Beekeepers have devised many ways of removing a box of honey-filled frames from the bees. When the moisture content of the comb has been reduced to around 18%, the honey cells are capped with beeswax. The bees leave an air gap under the capping and this can make the comb look very white. What you're looking for is a super where 75–100% of the comb surface is capped over, in order to remove it and extract the honey. You can wait until all the frames in the super have been capped sufficiently or you can take individual capped frames from different supers and 'make up' a super for extraction from several colonies. Bees in supers are quite tolerant of each other but you can shake the bees off each frame as you remove it from its colony. Remember – don't leave any gaps or the bees will fill them with wild comb. If you don't have frames of drawn comb, add frames of foundation at the outside edges of the super.

If not all of the cells in a comb are capped, hold it horizontally over the hive and give it a sharp shake. If nectar flies out of the cells then the honey isn't 'ripe' and is likely to ferment if you extract it. Put the frame back and wait for the bees to reduce the moisture content. However, if only a few drops shake out this signifies that even though the cells may not be capped, they do in fact contain honey that can be safely extracted.

To be absolutely sure, you can measure the moisture content with a refractometer (see page 112).

Below: White cappings over cells of honey.

Clearing supers of bees

Supers can be cleared of bees using a Porter bee escape or one of the variety of clearer boards known collectively as 'Canadian escape boards'. All these devices are designed as one-way gates, allowing bees to pass through one way (from the super to the rest of the hive) but not to return.

Porter bee escape

The Porter escape fits into a hole cut in a clearer board or inner cover. If your board takes two escapes, the bees will clear more quickly. The escape consists of two parts that slide together. The top has a central hole down which the bees pass. The bottom section has two pairs of flexible but reasonably rigid copper or plastic 'springs', which are attached to the outside of the escape but come together at the other free ends, leaving a gap just less than a bee space. A bee can push through into the rest of the hive but the gap is too narrow for it to make the return journey.

Below: The Porter bee escape.

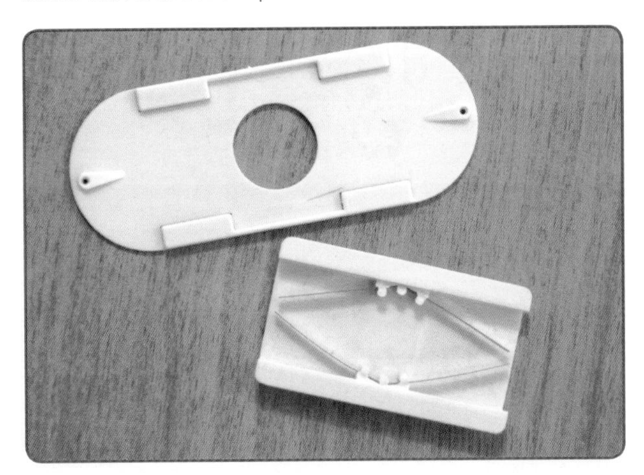

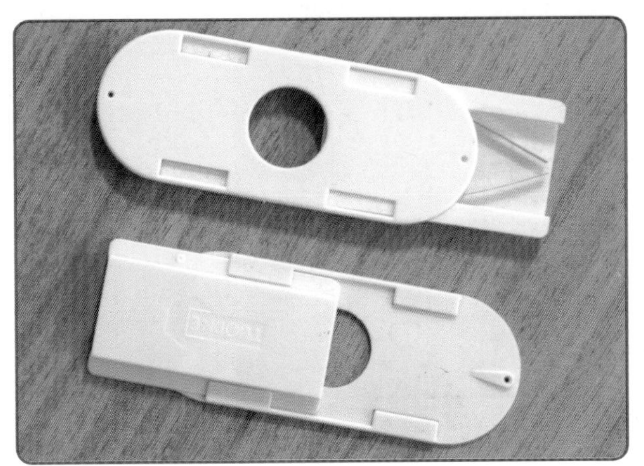

Canadian escape boards

These come in several designs based on the principle that bees can pass one way but are unable to return easily. One that works well is a lozenge-shaped plastic tray, roughly a bee-space deep, which is fixed under the clearer board. A hole in the board allows bees through and they're 'funnelled' to the openings at the ends of the lozenge. They don't return to the super unless the clearer board is left on the hive too long, when they can work out the way back. Holes in the lozenge allow the colony odour through, which attracts bees from the supers. I advise you to fix your Canadian escapes with drawing pins so that you can remove them easily to clean them. If they get blocked they will fail.

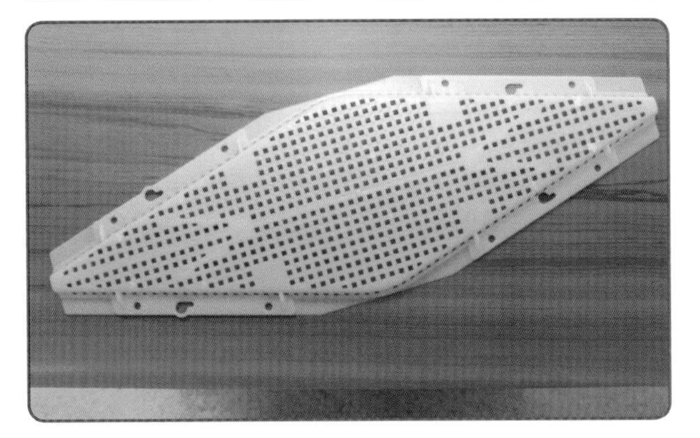

Above: A lozenge-shaped 'Canadian' escape.

Place your full super to one side. Return the other supers to the hive. Place the clearer board on top of the supers, the right way up with the exit hole uppermost. If you put the escape on upside down you'll get more bees than you know what to do with in the super you're trying to empty! Put the full super on top and add the inner cover and roof. Make sure that all the access points, such as the feed holes, are covered.

Good Canadian escape boards can empty a super of bees quite quickly. If you put one on in the morning, a super could be bee-free by late afternoon. Twenty-four hours should see 99–100% of the bees out, although

Above: The clearer board.

you'll probably find a few stragglers. Remove your cleared super and cover it to prevent bees gaining access while you reassemble the hive. Remove the clearer board and replace it with the inner cover as soon as possible. If you leave it on the hive the bees will start filling the gaps with wax and propolis, rendering the escapes useless. They're particularly fond of propolising the attached end of the springs in Porter bee escapes, which fixes them rigidly in position, disrupting the size of the gap – and they're very fiddly to clean.

Honey on the hive is warm because of the general hive temperature and it will flow more easily, so don't leave cleared supers around to cool down. While you're waiting for the clearer board to work you should be preparing your honey extraction area and equipment. Once you've removed the super, take it away to be extracted.

Ripe honey

Capped honey is said to be 'ripe'. It will have a moisture level of around 18%. In prolonged damp weather this may well be higher because honey is hygroscopic, absorbing moisture from the atmosphere. You can measure the moisture content of honey using a refractometer. These used to be expensive but cheaper models are now becoming available which are perfectly suitable for checking your honey. Why should you consider doing this? Firstly, because there's a legal requirement that the moisture level in honey that's sold must be below 20%. Secondly, if the moisture content is too high the naturally occurring sugar-tolerant yeasts will start off fermentation. This is good if you want to make mead, but not if you want to use or sell honey. So keep your honey covered with an airtight lid to prevent it absorbing atmospheric water, with the associated risk of fermentation.

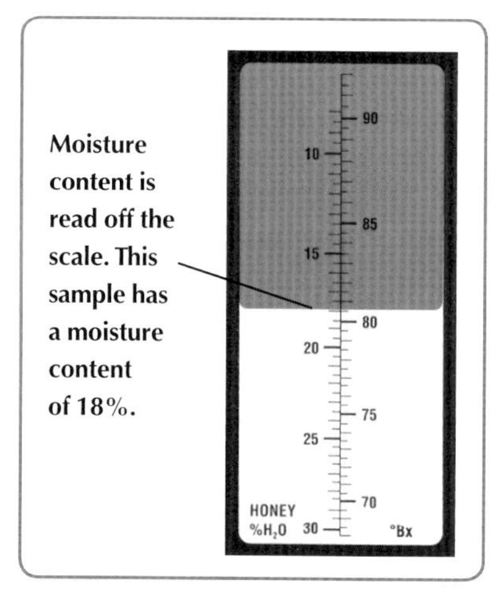

Right: The refractometer scale indicates moisture content in the honey.

Left: A refractometer.

The end of the season

At the end of the honey season, remove the remaining supers from your colonies. When you're just beginning you can get an idea of when to do this by talking to local beekeepers who have experience of the local honey flows and can advise you. After July, although the weather may be warm and sunny and the bees may be active, there is little forage that cannot be fitted into the brood box. Italian bees may well need feeding as they have a tendency to use incoming nectar to raise quantities of brood rather than storing it.

The sudden reduction in hive volume when you take off the supers may well cause bees to hang outside the hive. Don't worry, they'll usually sort themselves out. It won't kill them.

Also at around this time, reduce colony entrances to a size easier for the colony to defend from robber bees and wasps. A slot around a bee-space high (5–6mm) and 100mm long is generally suitable, but this depends on the size of the colony and its determination to defend its nest. Reduce the slot but keep an eye on the hive. If wasps and robber bees are still managing to gain access, make the entrance smaller. You can take it down to one bee-space wide if necessary. If returning foragers have to queue to get inside, this is better than the colony losing all of its stores to robbers. More workers will also take up guard duty.

The size of wasp colonies is reducing by this time of year and the sugar reward adults receive when they feed the larvae is dwindling. They're therefore looking for other sources of sweetness. Ripe fruit and jam sandwiches attract them but so do the honey stores in a beehive. It's essential that your hives are bee-tight with the only way in being through the entrance. This gives the colony the best possible chance of repelling unwelcome visitors.

Above and below: Reducing the entrance to a width that the colony can easily defend.

Below: A guard bee stops a wasp entering the hive.

Extracting the honey harvest

In your first year you may not get a honey harvest, especially if you've started partway into the season with a nucleus. However, if you do have a surplus from a small number of colonies your problem will be the need for extracting equipment and suitable premises. The first you can buy, but it will be expensive, especially on top of your other start-up costs. Many beekeeping associations have extracting equipment available for loan or hire and this will see you through the first few years until you decide whether to buy your own. A rule of thumb for assessing your crop and deciding how many containers you'll need is that a 'full' super holds around 9kg (20lb) of honey.

Honey and wax

Warm honey is less viscous than cold honey and will come out of the cells more easily. Therefore don't leave supers hanging about for days before you extract them, but deal with them straight away. Honey still warm from the hive may filter quite easily, especially through a coarse filter.

Using unwired foundation

If you don't anticipate a large crop or have access to an extractor, one solution is to fit your super frames with thin unwired foundation. The resulting honeycomb is cut into pieces and everything is eaten. You need a cutting board, a grid on which to stand the cut comb to drain, and a knife or special cut-comb cutter. The latter produces pieces that will fit into the cut-comb boxes you can buy.

An alternative is to cut the comb from the frame, wrap it in a muslin cloth or a jelly bag, break it up and allow the honey to drain out. If you only have a small amount to deal with there's no reason why you cannot do this, but be warned – it's very messy!

My advice would be to collect the honey in large food-grade polythene containers, fit the lid, put the containers in a warm place and wait 24 hours. The wax particles and air foam will rise to the surface and you can skim them off. They're perfectly edible and can be added to the winter feed. The honey underneath will be pretty clear but if you want to sell it you need to filter it further.

Below: Thin unwired foundation should be used for cut comb honey.

Uncapping

First you need to uncap the honey cell with a sharp or serrated knife. One with a long slim blade of at least 200mm (8in) works best. A blade longer than the frame depth makes uncapping easier. You'll also need a clean plastic washing-up bowl and a length of wood some 40–50mm (1½–2in) wide and long enough to span the bowl and rest on the edge. Cut a notch in each end large enough to fit over the bowl's rim to hold it stable. Cut a recess in the opposite surface wide enough to take the end of the frame lug.

Place your bowl on a firm table covered with plastic sheet or newspaper to catch the drips. Put the support across the bowl and stand the frame lug in the recess. Then cut off the cappings, moving the knife either upwards or downwards – you'll work out which suits you best. However, be careful the knife doesn't slip if you cut upwards. Slice just under the cappings. Angle the frame so that the cappings fall away into the bowl as they're removed, so that they don't stick to the comb surface. At this stage you can even-up any irregular comb surfaces. Don't worry if this means cutting quite a bit off the cells and/or if you damage the comb. Sometimes lumps will be pulled off, even down to the midrib – the bees will repair the comb next year and you can recover the wax and honey from the bowl.

Rather than slicing off the cappings, you can also lift them using the tines of an uncapping fork. Alternatively you can scratch them open using the points. As you do so, jiggle the fork from side to side, leaving behind a trail of broken cappings. Honey will come out of the cells but you'll get small pieces of wax as well which will need to be filtered out later.

You can also remove the cappings with a hot-air paint stripper, even though this is not its intended use. Play the hot air quickly over the surface of the comb and the cappings will simply melt and disappear. The honey doesn't get hot, the comb is not damaged, and the frames can be put into the extractor straight away. Any uneven surfaces can be levelled off with a knife the following year before the combs go back to the bees.

Uncapping a frame

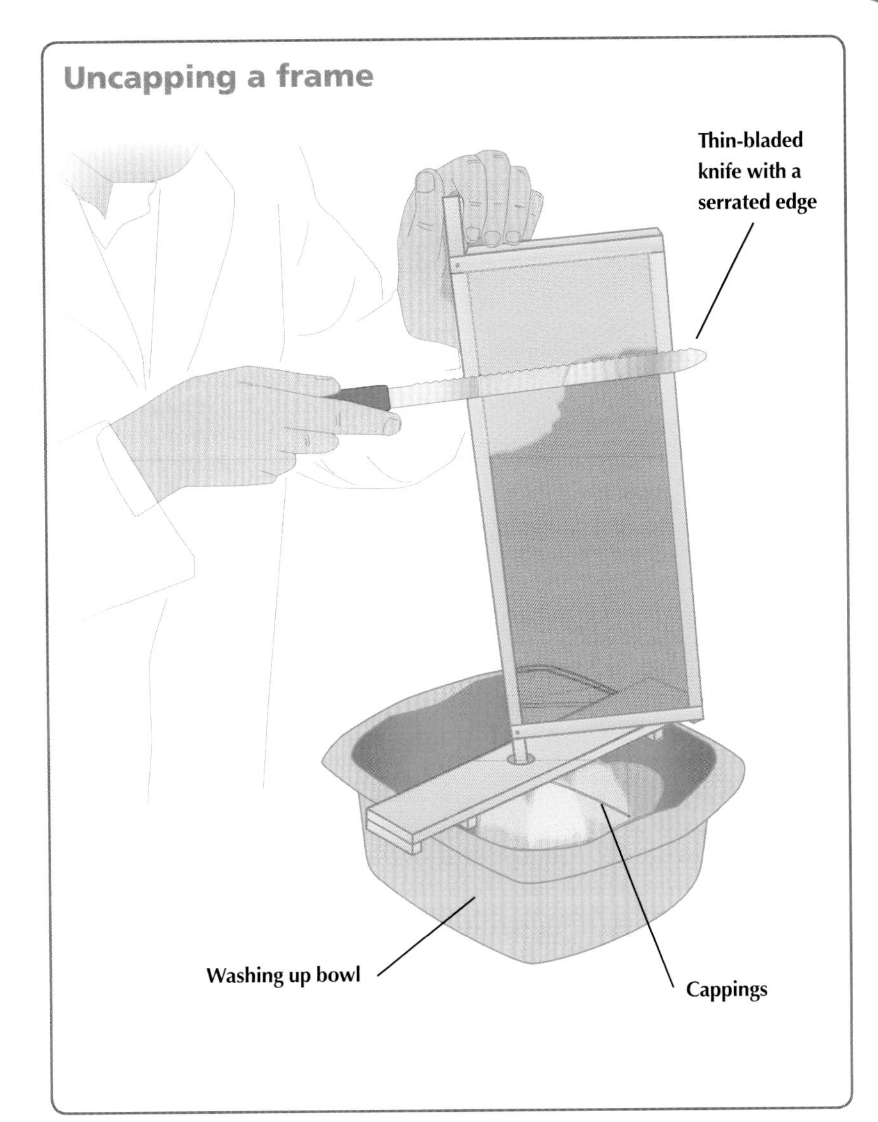

Below: An uncapping fork.

Honey extractors

Honey extractors come in two types: tangential and radial. Make sure that the one you choose will accommodate the frames you're using.

The tangential extractor

This has a rectangular cage that rotates on a central spindle inside the barrel. Frames are placed against the flat sides of the cage and liquid honey is thrown from the outer face only. The honey on the inside face stays put, so the frames have to then be turned vertically so that it can be extracted in the second stage. Frames should always be placed in the cage so that the top bar leads the direction of rotation.

Don't completely empty the cells on the outer face to start with because that leaves only soft wax supporting the honey in the inner comb. Rotate the cage until the outer cells are partly emptied. Turn the frame over, replace it in the cage and start to extract the honey from the other side. Repeat this as necessary until you've emptied both sides of the comb.

Above: A tangential extractor.

The honey that's thrown out hits the side wall and then runs down into a collecting well at the bottom. From here it's run off through a tap or 'honey gate' into jars or buckets.

Tangential extractors are usually small, holding two to six frames, and hand-powered, as you don't need to rotate the cage very quickly. If you turn the cage too fast combs can break up and not only will you have to filter the bits from the honey but you'll also have lost valuable drawn comb.

Below: The cage of a tangential extractor.

The radial extractor

The circular cage in a radial extractor holds the frames along the radii, like the spokes of a wheel, with the top bars to the outside. As the cage rotates the top bars move faster than the bottom bars and suction lifts the honey out of the cells on both sides of the frame at once. It's whisked away by centrifugal force and runs down the side walls into the collecting well.

Below: A radial extractor.

Radial extractors come in a range of sizes but most small-scale beekeepers will use one that takes nine, fifteen or twenty frames. They can be hand- or motor-powered. If yours is motor-driven, build up the speed gradually so that the extractor can get 'balanced'.

Below: Frames in the cage of a radial extractor.

Loading an extractor

Try to balance the frames in the extractor – that is, load ones of similar weight in a batch or opposite each other in the case of a radial extractor. When you start rotating the cage the extractor will begin to move around on its legs until some of the honey is extracted and the weight inside is evenly balanced. Then it will settle down.

Many years ago I was given a tip which I was convinced wouldn't work – until I tried it. When the extractor starts wobbling, your first reaction is to grab it and try to hold it still. Indeed, I've known beekeepers screw the feet to the floor. But this puts a great strain on the barrel, which is trying to move in the opposite direction from the cage. The advice I was given removes this strain.

Take some fairly substantial timber and fix it into a 'T' shape. The 'spokes' of the 'T' need to stick out a bit further than the ends of the legs when the extractor stands on them. Get some strong castors and attach one to the end of each spoke of the 'T'. Turn the 'T' over and screw the bottom of each of the extractor legs to one of the three spokes. Then, when you start rotating the cage, allow the extractor to 'dance'. In due course the forces will balance and it will settle down to a gentle wiggle. Initially you may have to restrain the extractor – I once had to tie one to a table leg to stop it waltzing off down the room!

A method of reducing the strain on the extractor barrel

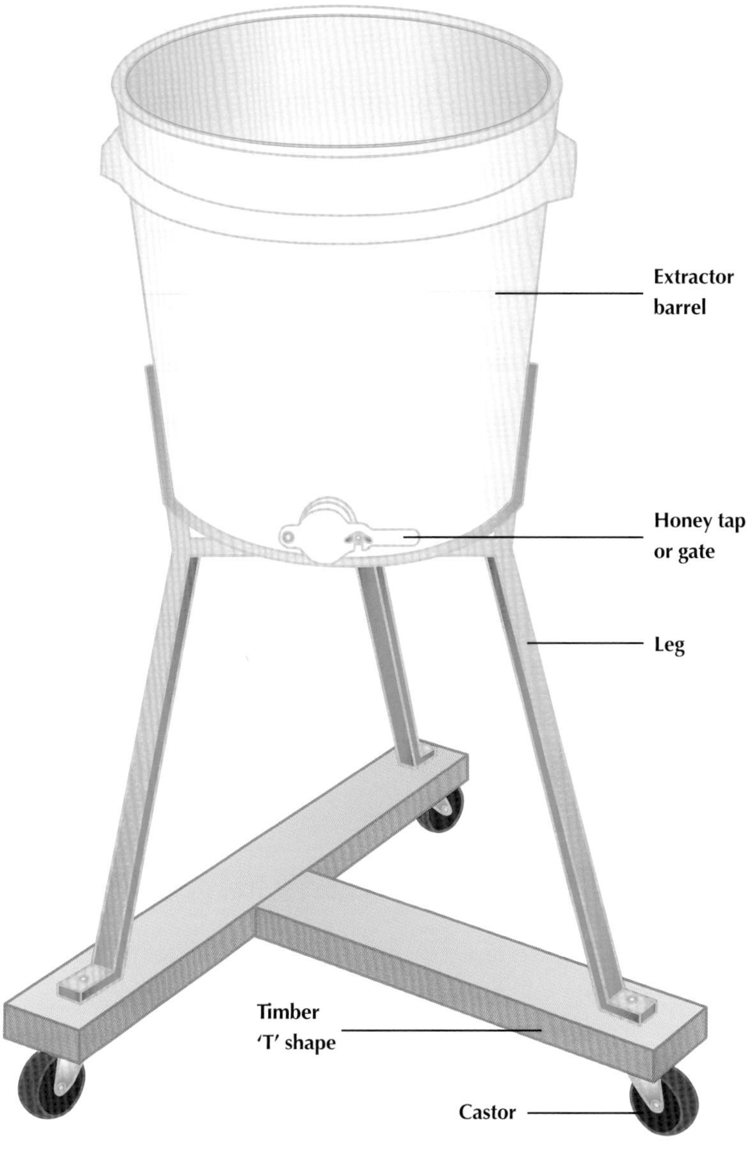

Emptying an extractor

Don't be tempted to run the extractor too fast or for too long. Getting out the last drops of honey will be at the expense of adding fine air bubbles to the liquid.

Remove the frames from the extractor and return them to the super. When you've finished, you can take the 'wet' supers back to the apiary in the evening and place them on top of the hives, over the inner cover with the feed hole open and the roof on top. Bees will get very excited at this sudden bounty but this will be reduced at night. By the morning things will have calmed down and the risk of robbing will have greatly reduced. Within a week all the remaining drops of honey will have been removed by the bees. If there are still bees in the supers they can be removed using a Porter bee escape in the feed hole. When the supers are cleared you can remove them for winter storage.

Honey in the well at the bottom of the extractor should be run off through the tap before it reaches a level that fouls the bottom of the cage. You can either filter and bottle it or run it into buckets and deal with it later. But be warned. Don't leave honey running out of a tap unattended. It flows silently and most beekeepers have a tale to tell of how honey got all over the floor. I knew a beekeeper in this situation who had two Airedale terriers who obliged by licking it up!

The extracting room

Honey is a food and, as such, if you want to sell it it's subject to legislation. This can be amended from time to time, so it's best to check on the Internet for the most recent version. The latest details are available on the website of the Food Standards Agency at www.food.gov.uk.

You should extract your honey in clean and hygienic conditions. For a small-scale operation, a domestic kitchen is suitable. You must have two sources of water available, one for washing hands and equipment, the other for washing the floor. I find it easiest to have water in a bowl for the floor and wash everything else in the sink.

To make life easier, arrange your equipment so that you can move easily from one task to another. Try to move dripping combs between stages over the equipment. As your hands and equipment get sticky, wash them before you start to spread honey over everything you touch. Wipe honey from the floor with a just damp rather than a very wet cloth (which actually spreads it into a larger sticky patch).

Try to arrange the stages in a circle around where you'll stand. Start with the sink and then have the pile of full supers. Next comes your uncapping station, which is next to the extractor. After the extractor you can pile up the empty supers that receive the empty frames.

Layout of the honey extracting area

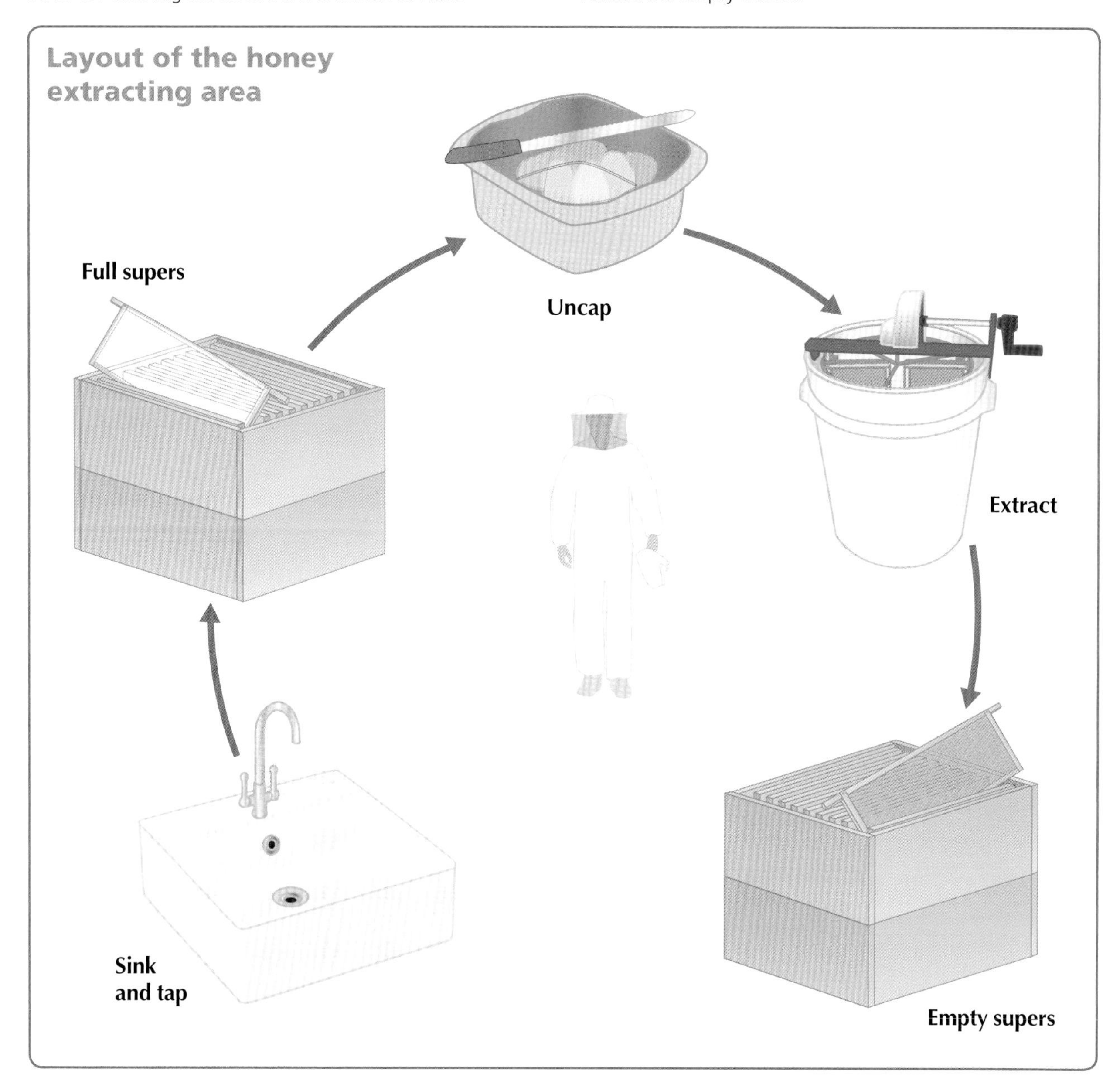

Filtering honey

If you plan to bottle your crop straight away you can filter it as it flows from the extractor using a conical tap strainer. This is hung from the tap, allowing the honey to pass through and into jars. Take care that you close the tap in advance of filling the jar so that it doesn't overflow.

Above: A conical tap strainer

Your filter will get clogged with bits of wax and you'll need to stop and clean it at regular intervals. Do this before the filtering process slows to a crawl or you'll be waiting a long time for the filter to clear so that you can remove and deal with it.

Below: A double honey strainer.

EH Thorne (Beehives) Ltd

Alternatively you can run the honey through a coarse filter into a 30lb (13.5kg) food-grade bucket. A double strainer with a coarse filter sitting on top of a finer one works well. The top filter catches the larger pieces of wax and the lower one removes finer particles. The honey will be fine for home consumption but if you want to sell it you need to filter it further.

The settling tank

First transfer your honey to a settling or bottling tank. This is simply a large food-grade vessel that will hold a good amount of honey. Those commercially available come in plastic or stainless steel and are designed to hold anything from 35kg (77lb) to 100kg (220lb).

Set your tank on a table so that you can operate the tap comfortably when seated and have somewhere within reach for the empty and full jars. Attach a fine filtering cloth across the top of the tank. I would advise you to allow the cloth to hang inside, even down to the bottom, rather than making it taut. You want to pour your honey slowly into the tank in a way that traps as little air as possible. One way is to run it down the inner surface. A loose cloth will enable you to do this. If the level of the honey is above the bottom of the cloth, simply lift the cloth and allow honey to pass through. You may have to stand on a chair to pour in the honey or you can do this with the tank on the floor and then lift it on to the table. It will be heavy.

Once the honey is in the tank, leave it for 24 hours to allow air bubbles to rise to the surface. You can then fill your honey jars from the tap. The first jar or two may contain small bits of wax or dirt so put these to one side for home consumption. Stop filling jars when air scum starts coming out of the tap. You can run the remainder into a bucket and use it yourself.

Below: A settling tank.

Honey jars

Although UK weights went metric in 1995, several items, including honey, strawberries and Christmas puddings, can still be sold in units based on imperial weights. However, the packaging must state the equivalent metric weight. The standard honey jar holds 1lb of honey but it must be labelled as containing 454g. The Imperial weight can be included on the label.

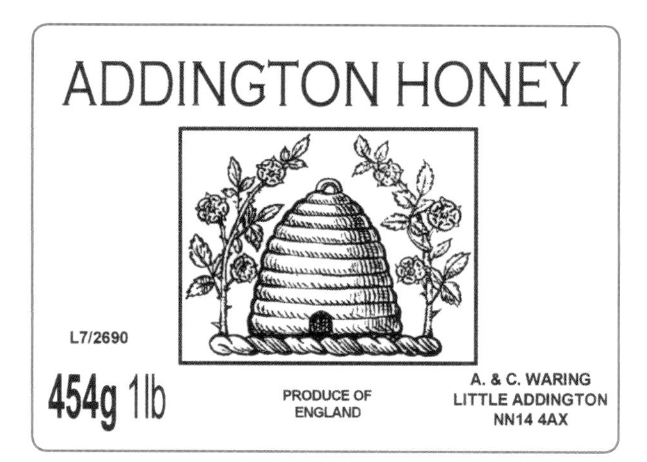

Above: Honey labels must conform to the regulations.

You may prefer to collect your jars from the supplier, as no responsibility will be taken for breakages during carriage. Check if your local association buys honey jars in bulk. These are usually at an advantageous price. Jars come with metal or plastic caps, which are often in place. These are clean and ready for use. If the caps come separately you should wash the jars before use. We bought our dishwasher primarily for this purpose as it sterilises and dries them. We then add the caps and store them until needed. The same goes for returns from satisfied customers. With returns, we wash and reuse plastic caps, but metal caps get scratched easily and the acid in the honey causes them to corrode, so we always dispose of them and use new ones.

Warming cabinets

You can invest in or make a warming cabinet to warm honey to, say, 35°C (95°F), at which temperature it will filter easily. This is the brood nest temperature so it's quite safe and will not spoil your honey. You can even use this cabinet to liquefy granulated honey but for this you'll have to set the temperature at around 42°C (108°F). The process may take 24–48 hours and must be thermostatically controlled. Don't heat honey to a higher temperature or for longer than it takes to liquefy it. The wax should remain intact, as its melting point is around 63°C (145°F).

HMF and diastase

In honey legislation, two substances have been chosen as indicators of the way it has been treated. One is hydroxymethylfurfural, or HMF, and the other is the enzyme diastase.

HMF levels in honey rise when it's heated or kept for a long period. The Codex Alimentarius international standard requires honey to have an HMF level below 40mg/kg to guarantee that it has not undergone heating during processing. Fresh honey has HMF levels of less than 15mg/kg, depending on pH value, temperature and age. If you sell your honey you're declaring that it conforms with these regulations.

Diastase is an enzyme introduced into honey by bees. It converts starch into the sugar maltose. Its level falls with temperature so it's another indicator of overheating. Currently the UK specification is not less than 8 on the Schade scale.

If you don't overheat your honey or keep it for a very long time or in warm conditions you don't need to worry about either HMF or diastase.

Granulation

Honey is a complex mixture of sugars with traces of other substances. The main sugars are fructose and glucose with the higher proportion being fructose. However, when there's a high level of glucose this crystallises from the solution, causing granulation. The rate of granulation depends on the glucose level and varies according to the source of the nectar. Honey from oilseed rape has a high glucose content and can granulate in the comb, so you need to remove and extract it while it's still liquid. You may have to extract it between mid-May and the end of June.

Below: Granulated honey showing some 'frosting' caused by trapped air.

Fermentation

Honey is most likely to ferment when it has crystallised. Granulated honey is composed of crystals surrounded by liquid honey. If exposed to the air, honey attracts moisture and the liquid honey then becomes more dilute and sugar-tolerant yeasts become active. This leads to fermentation and your honey will smell like ripe apples and taste faintly alcoholic. You can make sure this doesn't happen by only extracting honey which is ripe, in other words has a moisture content of less than 18%, and storing it in tightly closed containers. I have some honey that's 27 years old and shows no sign of fermentation at all. It probably has a very high HMF level by now but it still tastes fine.

Labelling

If you sell your honey in the UK you must comply with the Honey Regulations 2003. The second version was published in July 2005 and amended in October 2007. These regulations, and the Food Labelling Regulations 1996, give details of the type and composition your honey must be and how jars must be labelled. Rather than go into the details here, I recommend you start by purchasing labels from an equipment supplier, which will conform to these regulations.

Honey for sale must carry a 'Best Before' date and most beekeepers use one of one or two years. You must also include a Lot Number, beginning with L, on your jars and keep records of your batches and where they're sold. You can download guidelines from the Food Standards Agency at www.food.gov.uk, and a leaflet from the British Beekeepers' Association can be found at www.britishbee.org.uk/files/selling_honey_B10.pdf.

Above and below: Jars of honey on the show bench labelled, ready for sale, showing a wide variety of labels.

Uses of honey, beeswax and propolis

Most people start keeping bees to produce their own honey and there's no doubt that this is a very exciting experience. I can still remember taking my first super round to a beekeeping friend to use his extractor, and the thrill of taking home my first honey crop. In fact I still have some in the cupboard, proving that with the correct water content and stored in an airtight container and cool conditions honey can last without fermenting for a very long time.

Above: Honey cake

Honey

Honey is sweet but, unlike sugar, it has aromas and flavours, some of which can be very distinctive, such as that of ling heather honey. In the end it comes down to personal preference.

Honey is a mixture of sugars plus essential oils, traces of minerals and water. Typically, honey contains 17–18% water, 39–42% fructose, 34–35% glucose and 1–2% sucrose. The remainder is made up of complex sugars, proteins, wax, salts, acids, volatile oils, pollen grains and pigments.

As well as direct consumption, honey can be used in recipes as a substitute for sugar, provided you take its water content into account and adjust the recipe accordingly. For calculation purposes, it's near enough 20% water. Honey can also be used as the sugar required to start yeast working for bread-making, or for a glaze on roasted meat. But don't brush or pour it on too soon or it will burn.

Below: Different coloured honey from different nectar sources.

Above: Meads from different honeys.

Mead and other honey drinks

Mead is an ancient alcoholic beverage based on fermented honey. To make a good mead you need more than honey and water, although this mixture will ferment. Add yeast and a yeast nutrient together with an acid such as citric acid. Leave the mixture in a warm place and when fermentation is complete, rack and bottle it. There are several excellent books that will give you advice on making mead. Most commercially available meads are very sweet but you can make it dry. A good dry mead can be excellent, especially if you chill it before consumption.

Mead makers also make other variations such as melomel (made with fruit juice), cyser (with apples) and pyment (with red or white grapes). If you're not a wine maker you might find it easier to start with one of these.

Try some mead, and if you like it you can make all sorts of variations using only a small quantity of honey.

Beeswax

As a beekeeper you can obtain beeswax from two sources – cappings, and by melting down used comb that's been taken from the hive and is no longer required. At first you won't have a lot of beeswax, but you can wash it and keep it in a sealed plastic bag until you've enough for your project. When you uncap your combs, make sure you save the wax. Cappings wax is probably the cleanest and purest you'll get.

Cleaning beeswax

Before you can use your wax you must clean it to remove the honey and bits of dirt. To drain the honey from your cappings, place them on a grid over a food-grade container. Honey is a food and, as such, comes under the food handling regulations. You can add the drained honey to that coming out of the extractor.

Wash beeswax with clean, soft water. Filtered rainwater is fine or you can use distilled water – beeswax reacts to the chemicals in tap water and you'll lose some of it in the scum. Don't melt beeswax in aluminium vessels as a similar chemical reaction will occur.

If you only have a small quantity of cappings you can melt them in a microwave. Put some clean, soft water into a glass bowl and add the washed cappings. Turn the microwave to high and heat the cappings in short bursts. You don't need to boil the water as wax melts at 63°C (145°F). Boiling could produce foaming which will make a real mess if its splashes or overflows the bowl. You can also melt wax in a stainless steel pan, again with some clean soft water in the bottom. Again, don't boil it.

Remove the bowl from the heat, cover it with insulating material and let it cool slowly. The wax will solidify into a disc on top of the water and can be removed. There are two types of dirt in wax – sinking dirt and floating dirt. Both can be scraped off the wax surface.

Below: Melting beeswax over soft water.

Filtering beeswax

To clean your wax, you need to filter it when it's liquid. Your filter material will become caked with wax that will be impossible to remove, but you'll be able to use it more than once as the wax will re-melt when it becomes hot from the next batch. Eventually it will become too clogged to be useful and you can then roll it up and use it for firelighters.

A 'furry' material makes a successful wax filter and many beekeepers entering wax into a show, where it has to be perfectly clean, will use surgical lint, furry side upwards. However, this is expensive. Other materials that work well include old sheets and nappy liners.

Support your filter material firmly before pouring in the molten wax. One way is to line an old sieve with the filter. You can also open both ends of a clean tin, stretch the filter material over one end and secure it tightly with a rubber band or similar. Turn the tin upside down with the filter at the bottom and support it above the collecting bowl, the surface of which has been covered in a thin film of a releasing agent, such as washing-up liquid. Carefully pour the liquid wax into the tin, making sure it doesn't overflow. Add more wax as the level drops.

If you've planned ahead you can use your liquid beeswax straight away for your project. However, to save it for later simply pour it into moulds and let it set. Plastic containers used for soft butter or margarine are ideal. The wax will shrink away from the surface as it solidifies and the handy blocks can be removed. Wrap them in a strong plastic bag until you need them.

Above: Beeswax blocks.

Uses of beeswax

Beeswax can be used in a wide variety of applications including candles, furniture polish, soaps and cosmetics. Other applications include waxing pins for lace making, lost wax casting, waxing bow strings, encaustic art and making mouthpieces for didgeridoos.

Below: A wax filter can.

Below: Beeswax candles.

Above: Realistic carnations made from beeswax

There's not space to go into the details of all these ways of using beeswax. If you want to pursue any of them, I recommend looking for a detailed craft book or attending a specialised course. I'm sure you'll find it fascinating and very rewarding. There's something special about burning a candle that you've made with beeswax harvested from your own bees.

If you have enough you can use your wax to make foundation. Presses are available that will make sheets with the honeycomb pattern embossed and offset on both sides. Bees take readily to home-made foundation, but your sheets are thicker than those made commercially and you'll not get as many sheets to the kilo (pound).

Do not make foundation with beeswax recovered from colonies with American or European foul brood. AFB spores can survive the temperature of molten wax and you'll simply spread the disease throughout your apiary. Any such comb *must* be destroyed.

If you don't want to make your own foundation, you can exchange your filtered wax for new sheets at several of the larger beekeeping equipment suppliers. An association I know collects wax from its members and does a bulk deal with a supplier, members receiving the equivalent number of sheets of new foundation.

Above: A propolis screen.

Propolis

Propolis has antiseptic properties and these are utilised by several health product companies to make throat lozenges and propolis tablets or syrups. Plastic screens are available which can be placed on top of the hive. The bees then fill the gaps with propolis. The screen is placed in the freezer and when the propolis is cold it can be pushed out of the gaps by flexing the screen. You may be able to sell the propolis to the health product manufacturers, but check with them first.

Below: A range of propolis products.

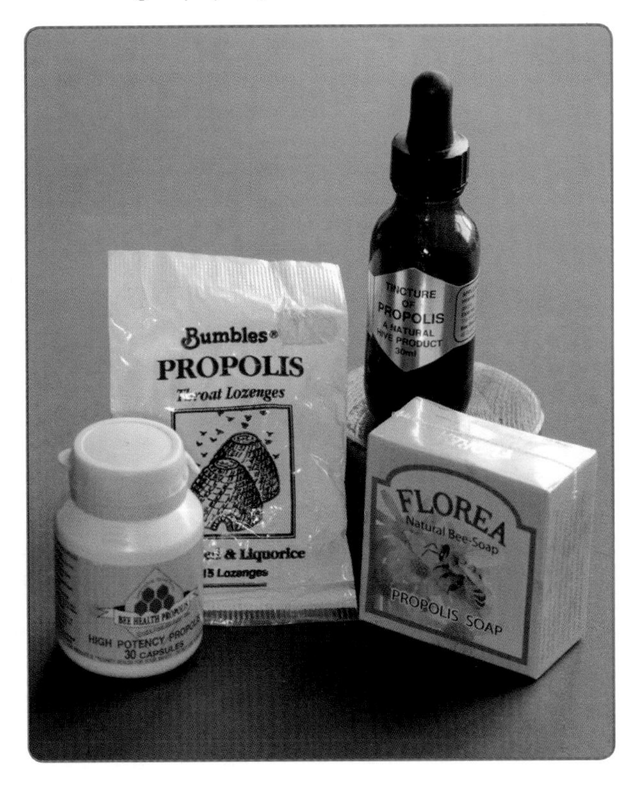

CHAPTER 5 BEE MANUAL

PREPARING FOR WINTER

Feeding

One of the most important reserves a colony has to try to maintain is food. There are times in the beekeeping year when this reserve is naturally low, such as in the spring, and naturally quite high, such as in autumn, just after the honey season. If you take some of these reserves it's your duty to replenish them. The bees collect them because they *need* them.

The best food for bees is undoubtedly honey but they'll readily accept and live on various sugar solutions. In Britain, the main feed is cane or beet sugar and untold numbers of colonies have, are, and will continue to survive on this at any time of the year.

It has been said that you should only feed bees when their reserves fall below a minimum level of about 4.5kg (10lb), which is roughly equivalent to two solidly filled brood frames of honey. However, you have to consider the whole picture. As you look through colonies it's quite normal for there to be very little honey in the brood box and hundreds of pounds in the supers. If you remove the supers for extraction, say in June, and the weather turns bad, your colony can starve to death. Feeding can, and should, be undertaken when and if the bees need it. It's quite possible with the right bees and management not to have to feed at all but this is something that you'll learn through experience.

It has been said that bees survive the winter better on sugar stores than honey. Bees will drink sucrose and may partly alter its make-up through a process called inversion. Honey contains nitrogenous matter such as pollen and this waste builds up in the bee's rectum during prolonged periods of confinement to the hive and, in species that aren't adapted to such conditions, makes it prone to dysentery. Inverted sugar syrups at the correct strengths to feed to bees are available and you might like to consider using these. Weigh up the costs of obtaining inverted sugar syrup against those of sugar and the work and mess involved in producing sugar syrup.

Making syrup

There's a great deal of confusion about how to achieve the correct concentration of sugar syrup. The syrup fed in winter needs to be more concentrated than that fed in spring or summer, as the bees have to work hard to evaporate the water to make the syrup suitable for storing.

The traditional winter feed consisted of 2lb of sugar in one pint of water. When metrication arrived, confusion reigned. However, there's a simple solution to making winter-strength syrup:

1. Put your dry measure of sugar (be it 1kg, 1lb or whatever) into a vessel that will hold it easily.
2. Roughly smooth the surface and mark the level.
3. Pour in hot water, stirring as you do so.
4. Bring the level to around 12–25mm (½–1in) above that of the dry sugar.
5. Stir until the sugar is dissolved (a few crystals left in the bottom don't matter).
6. If you're using a saucepan, gently heating the solution helps the sugar dissolve.

Occasionally you'll need to feed more dilute syrup, for example if your bees are running short in the spring/summer or if you're feeding a nucleus. If that's the case:

1. Put your dry sugar in the vessel. It should not come more than halfway up.
2. Mark the level.
3. Add hot water to double the volume.
4. Stir until the sugar is dissolved.

Some beekeepers get very precious about the correct syrup concentration. Personally I don't think it's that precise a science. The idea is to produce a syrup strong enough to require the least amount of work before the bees can store it safely. Bees collect nectar of various sugar concentrations and make it all into honey. Just feed them. They'll sort it out.

Feeders

The syrup is offered to your bees by means of feeders placed on the hive. These come in different shapes and sizes but basically consist of a well for the syrup and a means of access for the bees.

'Contact' feeders

A contact feeder does exactly what it says, bringing the sugar syrup into direct contact with the bees but delivering it in a controlled manner.

They can be as small as a honey jar or as large as a 30lb honey bucket. They're the only type that can be used to give food to bees under all conditions. In winter they can be placed directly on the top bars over the winter cluster and 'contact' the bees. The winter cluster cannot break up to use a larger 'rapid' feeder.

Contact feeders work on the principle of the partial vacuum caused by air pressure. To understand how they operate, take a used jam jar and fill it with water. Use a one-inch round wire nail or a skewer to make three or four holes

Right: Holes punched in the lid of the jar.

Above: Placing the feeder over the cluster. It is then surrounded by a super or eke, an inner cover and the roof.

Right: Inverting the jar and catching the drips of syrup.

Right: A partial vacuum holds the syrup inside the jar.

You can use a larger contact feeder, such as a honey bucket, to give your bees their winter feed but this will need something like a brood box round it to prevent robbing.

Above and below: A purpose-made contact feeder with a mesh area from which the bees can drink the syrup.

in the lid. Tighten the lid on the jar and turn it upside down over the sink. A little water will run out and then stop. A partial vacuum has been formed in the jar and air pressure holds the liquid in.

To feed your bees in the winter because of some emergency, fill the feeder with 'strong' syrup (a 1:1 sugar:water ratio), upturn it over a bucket to catch the drips then place it on the top bars directly over the bees. You may have to support it on two or three small pieces of wood to give the bees room to move around underneath the lid. Prevent robbers gaining access to the feeder by placing an empty super around it to keep everything bee-tight. Cover the feeder and frames with a sheet of plastic and then some insulation (newspaper works well) to help keep the syrup warm for as long as possible. Then add the inner cover and the roof.

'Rapid' feeders

The most common rapid feeders are the round type and the type that covers the whole of the top of the brood box. They work on the same principle: there is a well for the syrup and a gap in the centre or at the side that allows bees to climb up from the hive. A barrier prevents them reaching the syrup, which seeps from the well through a narrow gap at the bottom.

ROUND FEEDERS

Round rapid feeders are placed over the feed hole. The bees climb up and over the central column to access the syrup. The column is covered by a larger cylinder that confines the bees to the narrow gap between the two. The outer cylinder often has a clear top so that you can see if the bees are there. The bottom edge of the outer cylinder doesn't fit flush but has a means of allowing the syrup into the gap.

I suggest you fill the feeder with water in your house first to determine and mark the 'safe' level. Take the feeder to the hive empty and fill it with syrup from a carry-can. Afterwards it can be refilled as often as required, by simply lifting off the lid.

Round feeders need an appropriate surround, such as a super on which the roof goes, to keep robbers out.

Below: A rapid feeder.

'ASHFORTH' OR 'MILLER' FEEDERS

These feeders follow the same basic principle but they fit the cross-section of the hives for which they were designed. They cover the whole cross section of the hive and are flush with the outside of the brood box. They're usually made of wood, although plastic versions are available for some hive types. Unfortunately the wooden ones are prone to leak and each year it's worth filling any gaps and painting them inside before testing whether they leak. Putting a leaking feeder on a hive will certainly attract robbers!

The Miller feeder was probably the first to be devised. It has a central access slot with a chamber at each side for the syrup. The slot is covered by an inverted, square-angled,

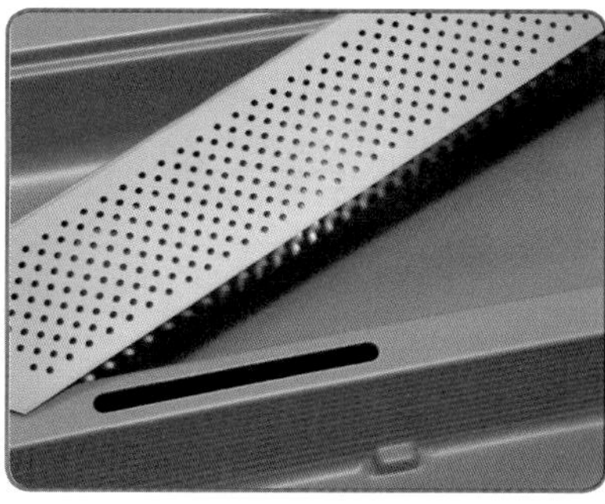

Above: A Miller feeder.

Left: The access channel.

U-shaped sheet of metal or plastic that fits over the inner walls of the syrup wells and covers the slot. This stops the bees from climbing over the wall and drowning in the syrup.

Since hives aren't always dead level, put this feeder on the hive with the slot running from the highest to the lowest side. The last syrup in the wells will drain to the bottom corners next to the slot and the bees should be able to collect it all. Placing the feeder so that one of the wells is on the lowest side of the hive means the last syrup will drain away from the slot. Each well should hold at least 4.5 litres (a gallon) of syrup. The level of the inner walls of the wells is lower than the depth of the feeder so don't overfill it or the syrup will pass over the top and drown the bees in the hive.

The Ashforth feeder was originally made from tin plate and sat inside a super. It was quickly modified to make it fit the hive like the Miller feeder, but with the covered slot at one side rather than in the middle. This means the whole of the feeder apart from the slot is available for syrup. Position an Ashforth feeder so that the end with the slot is at the lowest point of the hive. The bees can then empty it completely. In use, it's probably a better option than the Miller feeder.

It's not unknown for bees to ignore rapid feeders even when they're full, simply because they don't happen to find the syrup. So to make them aware of the feed you need to 'prime' the colony. When you've filled the feeder, pour a little syrup down the access point (cylinder or slot) on to the top bars to make sure the bees become aware of what's on offer. I mean about a tablespoonful – you don't want to drown your bees.

Frame feeders

These have the same outside dimensions as a brood frame and usually replace one when in use. Bees access the syrup through a gap in the top bar, and a float inside ensures they don't drown. Frame feeders can be put right next to the cluster but have the disadvantage that you have to dismantle the hive to refill them. Wooden ones can also leak. They cannot be used for emergency winter feeding as the bees cannot break the cluster to access the syrup.

Below: A frame feeder inside the brood box.

Emergency feeding

Winter

If you've fed your bees properly in the autumn you shouldn't need to feed them in the winter, but for one reason or another this is sometimes necessary. Traditionally candy was fed. This is made by boiling a sugar solution until it's thick enough to pour into moulds and set. An easier alternative is to find a baker who sells baker's fondant, which is like very stiff icing. Candy can be purchased from bee equipment suppliers.

Place the candy directly on the top bars over the cluster. Cover it with plastic to stop it drying out. To prevent robbing, surround the candy with an eke or super plus the inner cover and roof.

If your bees are without food and you have some of your own honey available, use this as long as you're sure your colonies are free of diseases like American and European foul brood. The bees would be eating their own liquid honey anyway and this is more easily digestible for them than candy, which they have to moisten with saliva to make a solution before they can drink it.

If you've run out of honey, *don't* buy a jar of foreign honey from the shop. You'll run the risk of introducing diseases into your colony, which is definitely not a good idea.

Bees have a very marked interest in complex scented sugar solutions. If you feed bees undiluted liquid honey, the extra concentration of 82% sugar means that they show a steady but controlled response. However, if you expose sugar syrup or honey in the open the bees will respond very readily. The moral is to make sure that when you're feeding bees you don't spill any!

Summer

Removing supers may leave a colony with no food stores, and a spell of poor summer weather may lead to starvation. If you feed syrup over empty supers on the hive they might store this in the supers. If you then extract these, could you claim to be selling honey? You could make candy and feed it to them but a simpler solution is to offer them granulated sugar in its bag.

Make a hole in the side of a bag of sugar to give the bees access. Left like this, bees are likely to carry many of the sugar grains out of the hive and drop them on the grass. The trick is to wet the sugar. You want the crystals to stick together so that the bees drink the sugary water rather than remove the crystals. Submerge the bag in a container of water until the sugar is wet. You can do this at the hive. Then place the bag on the inner cover with the hole near the feed hole. Add the usual super and roof to prevent robbing.

This is like us living on a diet of porridge. You wouldn't starve but you wouldn't look forward to dinner. However, the colony will be able to keep ticking over and it won't die of starvation.

You need to recognise when your colonies are short of food and understand about feeding, even at a very minimal level.

Below: Candy or fondant for emergency feeding.

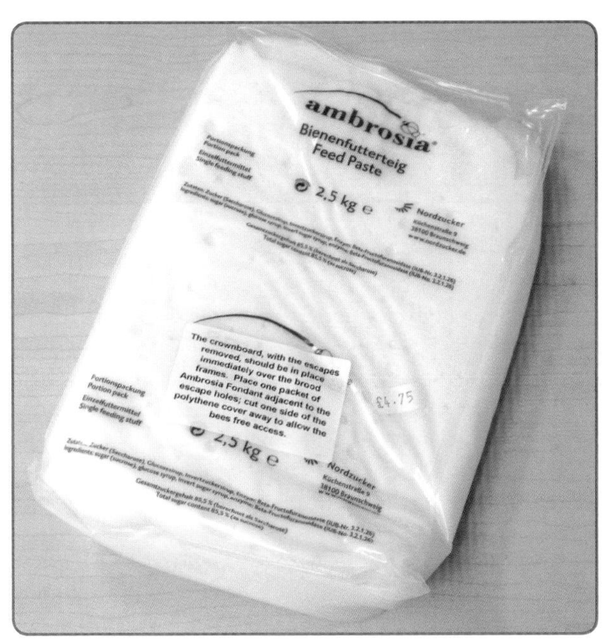

Preparing for winter

Honey bees in temperate climates have to survive a cold, foodless period – the winter. Because bees operate as a superorganism as well as individual insects, they cope with the cold by first creating warmth using their own bodies and then retaining this heat by crowding together in a cluster. Within the nest, the cluster takes shape when the air temperature falls to around 14°C (57°F). The size of the brood nest has already been reduced, so it can be covered easily. Bees keep the brood nest at a fairly constant 35°C (95°F). Any bee on the outside of the cluster that gets colder than 9°C (48°F) is unable to move and will die.

The winter cluster

The cluster is centred on the brood nest, or where the brood nest was. Broodless clusters have a lower core temperature of 21°C (70°F). Bees create warmth by rapidly contracting and relaxing their flight muscles, using stored food as the fuel. The cluster forms next to the stores and then moves upwards, keeping in contact with the food. If bees consume all the honey above them, the cluster can move sideways to other food stores. However, if these run out too the colony will die of 'isolation starvation'. Even if the food is only 20–30mm away from the edge of the cluster, the bees cannot move to it.

Below: The bees need food above the winter cluster.

Above: Guard bees attack a robber from another colony.

Feeding

Help your bees by making sure that they don't run short of food. Dark bees are well-fed with 20kg (44lb) of stores available. Other bees may need more. A British Standard brood comb full of honey contains about 2.3kg (5lb), or 1.15kg (2.5lb) per side. Go through the brood nest and estimate the amount of stores. Then 'heft' the hive: raise the back and side of the hive in turn just enough to slip an imaginary postcard between it and the stand. Try to relate how this feels to your estimate of the stores present. With practice you'll be able to estimate stores just by hefting. Then feed the appropriate amount of strong sugar syrup using your feeder of choice. A well-fed colony will feel as though it's nailed to the stand.

Robbing is best controlled by never letting it start. Don't spill syrup. Don't have leaky feeders. Make sure the only way into the hive is through the entrance. Reduce this to a size the guard bees can defend. If bees have to queue to get in, more of them will be stimulated to become guards.

'Winter bees'

Autumn flows from plants such as ivy, Michaelmas daisy and Himalayan balsam will supply bees with the pollen they need. Bees in town do well here as they have access to a great variety of pollen sources. They'll eat a lot and store the rest. This all helps to produce the so-called 'winter bees' about which we've already spoken (page 81). These have large 'fat' bodies and well-developed brood food glands,

Above: Ivy provides late pollen and nectar but do not rely on this for the winter feed.

and their lives are measured in months rather than weeks. Bees emerging in October/November could still be alive and foraging the following March.

Spring feeding

If any colony is short of food in the spring then the weaker syrup (1:1) described earlier is what's required. As explained earlier, don't feed this weak syrup to colonies for the winter as it contains a large proportion of water, and reducing this will wear out your winter bees. Only feed dilute/weak syrup when bees are able to fly from the hive at least once a week on average.

Water

Bees also need water to wet granulated stores with saliva so that they can feed. Healthy bees have no problem but they must have water in warm places near the apiary.

Damp and ventilation

Damp conditions are not good for bees. Hives must be weatherproof and at least 30cm (12in) off the ground to allow air circulation below the floor to keep it dry. However, hives placed in the full blast of an easterly wind will find survival more difficult than those in a sheltered spot. You need to make sure your apiary site protects your bees from wind in the winter but is sufficiently exposed so that the winter sun can touch the hives. This extra degree or two of warmth may be all that's required to enable the bees to fly and defecate. Ridding themselves of the waste products of digestion is very important and hives in the shade may well have to wait for a warm day.

The current belief is that bees need ventilation through the hive in spite of the fact that they'll seal all holes, cracks and crevices with propolis. The less aggressive the through draught, the better. To stop bees propolising things in late October/early November (in the Midlands – make allowances for your own geographical location) you can raise the inner

Left: Raising the inner cover to provide ventilation.

Left: Keeping out the mice.

cover 2–3mm (⅛ inch) by putting a matchstick or similar under each corner. You need to remove these by the end of February. If you have deep entrances, you need to fit a mouseguard at the same time and to protect your hives against woodpeckers if they're a nuisance in your area.

Spring losses

Most colonies that are likely to die begin to do so at the end of winter. The extra work imposed on bees in a small cluster or the effect of diseases like Nosema have their full effect when the bees are called upon to make the extra effort to rear brood in the spring. Poor spring weather and a shortage of pollen are the worst conditions for a colony. Common causes of colony death include starvation, queen failure, disease and being overturned by livestock or high winds. Generally, healthy bees with plenty of stores and a good queen will survive. You'll know that all is likely to be well when you see the winter cluster expanding and large pollen loads being taken into the hive.

Below: Pollen being taken into the hive.

PESTS AND DISEASES

Honey bee diseases

Mention of disease often fills a novice beekeeper with dread, but it's very important that you're familiar with those that may occur. By deciding to keep bees, you've accepted the responsibility to care for them in the best way possible, and as a beekeeper you must keep a check on your colonies and try to spot any signs of disease as early as possible.

Honey bees are susceptible to a number of species-specific ailments. These can be divided into those affecting the brood and those seen in the adults. However, the first step in successful disease control is to be able to recognise healthy brood.

Above: Empty cells over the wire in the foundation.

Healthy brood

First you need to recognise that the queen is laying a good brood pattern. This means you'll find brood of different ages in concentric circles on a brood comb. The queen will be laying in virtually every cell, so they'll contain an egg, a developing larva or they'll be sealed. Missed cells are probably easiest to see in areas of sealed brood.

Above and below: Healthy brood and a frame with a good brood pattern.

You may see straight rows of empty cells and wonder what's happening. Look into these cells and you'll see that they're the ones over the frame wire. Some queens simply don't like laying in such cells but this doesn't indicate a poor brood pattern.

Eggs

An egg should sit in the centre of the base of the cell. It will be curved over slightly and, looking over the comb, eggs should be in cells next to ones containing small larvae.

Above: Eggs stand upright at the bottom of the cell.

Below: Healthy larvae lie curled up in their cells.

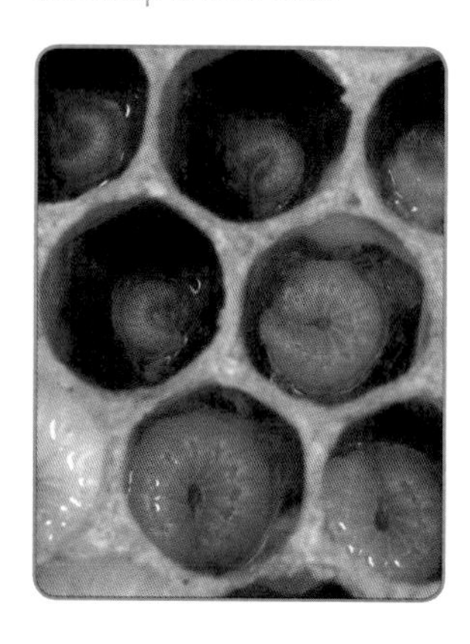

Larvae

These range in size from around that of the egg, when they've just hatched, to those that fill the cell. They should be pearly white and lie comfortably in the cell. You should find larvae of roughly the same age in curved patches as part of the normal brood pattern.

An egg takes three days to hatch. A developing worker larva is sealed into its cell six days later so you should find twice as many larvae as eggs if the queen is laying consistently at her peak.

Sealed brood or pupae

The dominant feature of an established brood nest should be the sealed brood. Since it takes 12 days for a worker larva to pupate and emerge as an adult, there will be twice as many sealed brood cells as there are larvae.

Above: Sealed brood dominates the brood nest.

If the comb has been used to rear several generations, the cappings will be dark brown. They're lighter on newer comb and just off-white on brand new comb. Healthy sealed worker brood will have uniform, slightly domed cappings. Cappings on healthy drone brood will have a more pronounced dome to accommodate the larger insect.

Left: Flat worker and domed drone cappings.

When an adult emerges, it leaves its cocoon behind lining the cell and covering the larval faeces. Worker bees clean and 'polish' cells before the queen re-lays in them. If your colony is free of disease, this is known as 'clean comb'.

Brood diseases

These fall into two groups. American foul brood (AFB) and European foul brood (EFB) are notifiable. If you suspect your bees are affected, then *by law* you must notify the authorities. The other common 'minor' brood diseases are chalkbrood, sac brood and bald brood. Bees are also susceptible to a number of viruses. Laying workers and a drone-laying queen are not strictly diseases but they do affect the brood and the colony.

American foul brood

This is not a geographical designation but, as with European foul brood, indicates where the disease was first discovered.

AFB is caused by the bacterium *Paenibacillus larvae* subspecies *larvae*. This develops in the bee's gut and kills it after the cell has been capped. Millions of spores are produced, which are ingested by house bees as they clean the brood cells. They then pass them on to new larvae in the brood food.

AFB spores are very tough, are resistant to extremes of heat and cold and are unaffected by disinfectants. This makes the disease very virulent. Spores can lie dormant for many years and then re-infect a colony. Eventually AFB will kill the colony.

A major method of spread is through the beekeeper moving combs, honey and equipment between colonies. I would advise you never to buy second-hand comb. If you purchase a colony of bees, make sure it's inspected by an experienced beekeeper to be certain that it does not have any disease, particularly AFB or EFB. In England and Wales your Seasonal Bee Inspector will check it for you if you don't have anyone else locally who can help.

SYMPTOMS

- Sunken, perforated cappings, which may look greasy.
- A 'pepper-pot' appearance where numerous cells have been uncapped randomly.
- A bad smell, although this is not definitive.

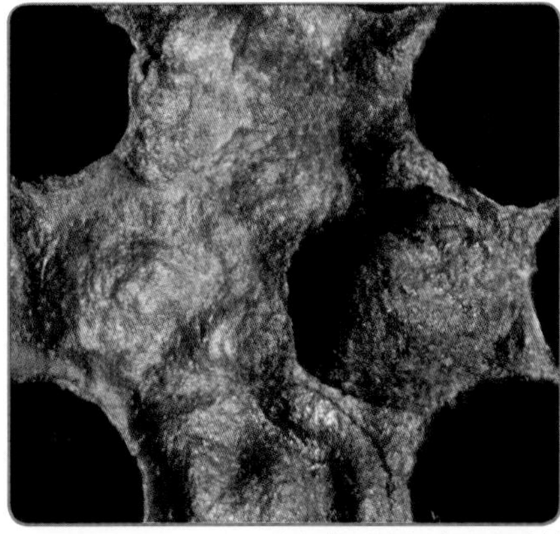

Left: Sunken cappings indicating American foul brood.

Below: The 'pepper-pot' patchy brood pattern.

There are two field tests for AFB. The first is the 'ropiness' test. Select what looks like an infected cell and open up the capping. Take a matchstick, poke it into the cell and stir up the larval remains. When you withdraw the matchstick, the cell contents will form a mucous-like 'rope' which can be drawn out to 10–30mm (½–1½in).

National Bee Unit, Fera (crown copyright)

Left: The 'ropiness' test for American foul brood.

The second test employs a lateral flow device, designed and marketed by Vita (Europe) Ltd. This works on the same principle as a pregnancy test kit. A larva with suspicious symptoms is extracted with a spatula and placed in an extraction bottle. This is shaken for 20 seconds and then a sample withdrawn with a pipette and a drop placed in the well of the device. After about 30 seconds a blue control line appears. The result can be read after another one to three minutes. A second blue line indicates AFB, no line means your colony is clear.

Below: The lateral flow device test kit.

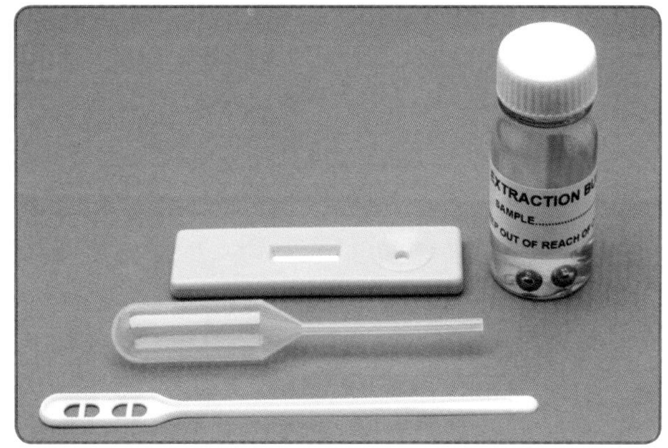

National Bee Unit, Fera (crown copyright)

National Bee Unit, Fera (crown copyright)

Left: A sample is taken of a suspect larva.

National Bee Unit, Fera (crown copyright)

Left: It is placed in the extraction bottle.

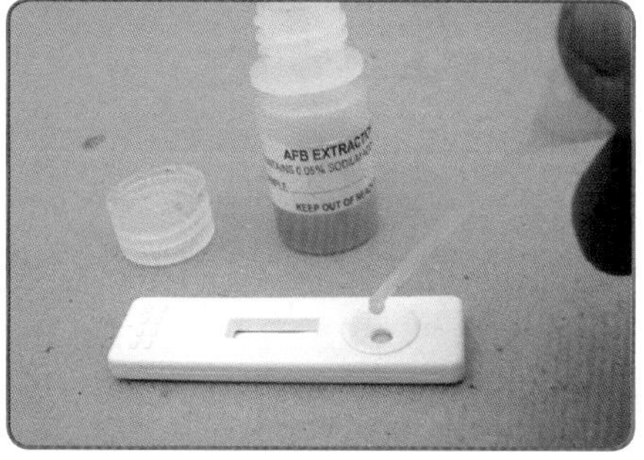

National Bee Unit, Fera (crown copyright)

Left: After shaking, a drop is placed in the well of the device.

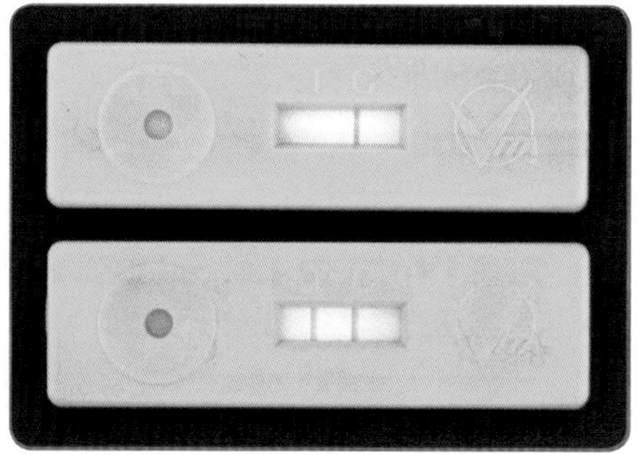

National Bee Unit, Fera (crown copyright)

Left: One line (top) indicates a clear test. Two lines (below) indicates disease.

The steps taken when using a lateral flow device

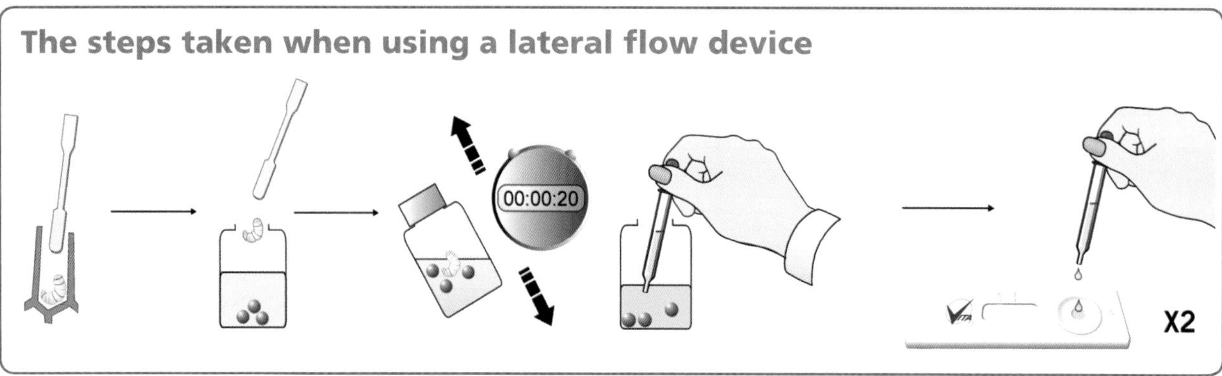

When the larval remains dry they form a very dark brown scale which the bees cannot easily remove from the lower surface of the cell. To inspect a comb for such remains, stand with a light source coming over your shoulder. Hold the frame at an angle of roughly 45° and look down into the cells. You'll see the scales on the lower surface.

National Bee Unit, Fera (crown copyright)

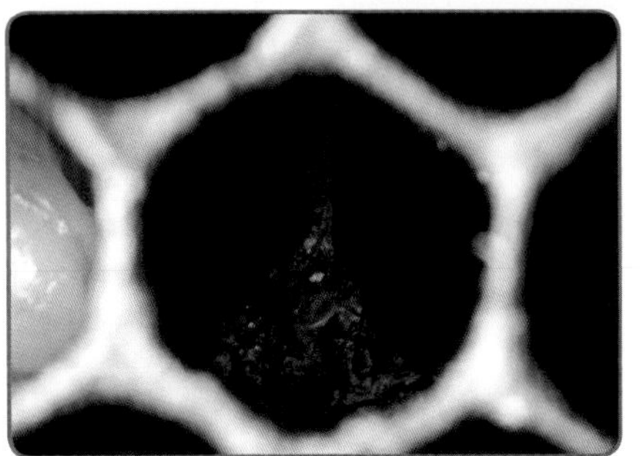
National Bee Unit, Fera (crown copyright)

Above and left: The dried scale of American foul brood.

CONTROL

The UK has no legally permitted treatment for AFB. If the disease is confirmed, the bees are destroyed. All the frames, combs, bees and honey are burnt in a deep hole and the remains buried to prevent robber bees visiting the site and taking spores home. Wooden hive parts are scorched to a chocolate brown colour with a blowtorch that produces a hot enough flame to kill the spores.

Below: Burning a colony with American foul brood.

National Bee Unit, Fera (crown copyright)

European foul brood

This is so called because it was first discovered in Europe. As with AFB, it's found worldwide.

EFB is caused by a bacterium, *Melissococcus plutonius*. This infects the gut of a developing larva, competing with it for food, and the larva starves to death shortly before the cell is due to be sealed. Not all larvae are killed by EFB. Some continue their development and emerge as adults. EFB is passed on when the bacteria are voided by the larva with the rectal waste, just before it starts to pupate.

After the larva dies, secondary bacterial infections often occur which can have a bad smell.

EFB can occur at any time of year but it's most obvious in the spring. It can be transmitted when bees rob an infected colony and it can also be carried by swarms. However, the beekeeper is once more the main culprit by moving infected combs, honey and equipment between colonies. This is another reason for not buying second-hand combs.

SYMPTOMS

- Infected larvae lie uncomfortably in awkward positions in the cells.
- Dead larvae can appear 'melted', and they turn yellowish-brown.

Right: Larvae with European foul brood lie uncomfortably and have a 'melted' appearance.

- With a severe infection large numbers of larvae are killed, giving a patchy brood pattern.
- There may be an unpleasant smell from secondary bacterial infections.
- Cappings over larvae that die in sealed cells appear sunken and may be perforated.
- The larval remains do not 'rope' as with AFB.
- Dried larval remains form scale that can be removed easily from the cells.
- EFB can be confirmed in the apiary with a lateral flow device.

Below: Sunken cappings from European foul brood.

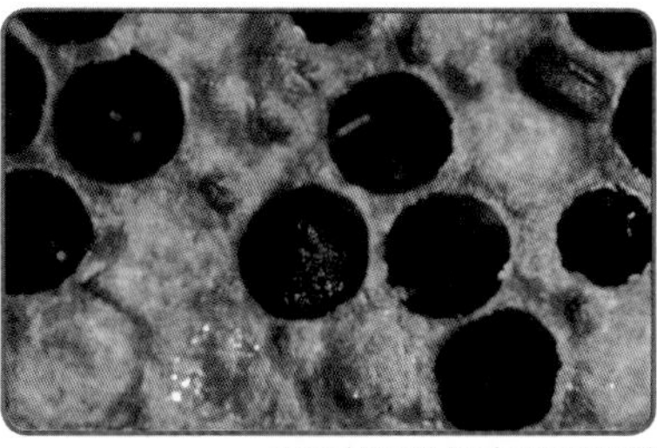
National Bee Unit, Fera (crown copyright)

CONTROL

Biotechnical control of EFB is achieved by a process known as shook swarming. Bees are shaken from their combs into a new sterilised hive containing frames of foundation. To carry out the 'shook swarm' technique you'll need:

- A clean brood box into which the colony is to be transferred.
- A second box for temporary storage of the frames.
- A full set of frames fitted with foundation.
- A queen excluder.
- A clean floor.
- A rapid feeder.
- 2.5 litres of 2:1 sugar syrup.

THE METHOD:

1 Scrape any wax and propolis from the new brood box and floor.

2 Flame the wooden parts with a blowtorch until they turn chocolate-brown, to destroy any disease spores.

3 Scrub metal parts with a strong solution of washing soda and rinse clean.

4 Move the colony to one side. Place the clean floor on the original site.

5 Place a queen excluder on top to prevent the queen leaving the hive.

6 Add a clean brood box filled with frames of foundation.

7 Remove three or four frames from the centre of the brood box to make a gap into which to shake the bees.

8 Find the queen, place her in a queen cage and put her in your pocket or somewhere safe.

9 Take each frame in turn, hold it well down in the gap in the new box and shake it sharply to dislodge the bees.

10 Brush any remaining bees into the new box.

11 Place the cleared comb into the spare box and cover it to prevent bees gaining access.

12 Knock or brush any bees clinging to the original brood box into the new one.

13 Shake or brush any bees on the original inner cover and floor into the new brood box.

14 Replace the frames of foundation previously removed (step 7) to fill up the gap. Rest them on the bees and then gently ease them into position as the bees climb up on to them.

15 Release the queen back into the colony.

16 Add a rapid feeder and fill it with 2.5 litres of strong sugar syrup (2:1 sugar:water) to help the bees draw out the foundation.

17 Replace the clean inner cover.

18 Replace the roof.

The colony will settle down and start drawing out the foundation. When the queen has started laying in this, remove the queen excluder from below the brood box.

A light infection can also be treated with the antibiotic oxytetracycline. This must be administered by the Seasonal Bee Inspector, who will explain the process to you. Severely infected colonies are destroyed and burnt in the same way as those with AFB.

Minor brood diseases

Chalkbrood

This is caused by a fungus, *Ascosphaera apis* which kills the larva after the cell has been sealed. The name derives from the larva's chalky white appearance as the fungus develops. Later on it may turn black. You may see these 'mummies' on the floor of the hive and near the entrance, as the bees can remove them from the cells.

Chalkbrood is spread by spores sticking to combs and adult bees. They can remain infectious for three years or more but chalkbrood is rarely a serious disease.

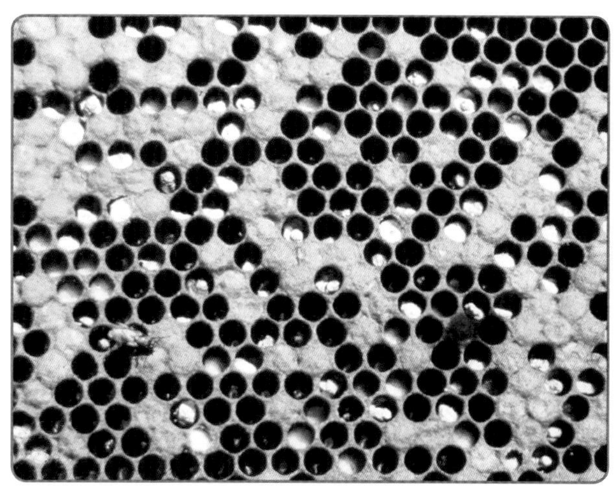

National Bee Unit, Fera (crown copyright)

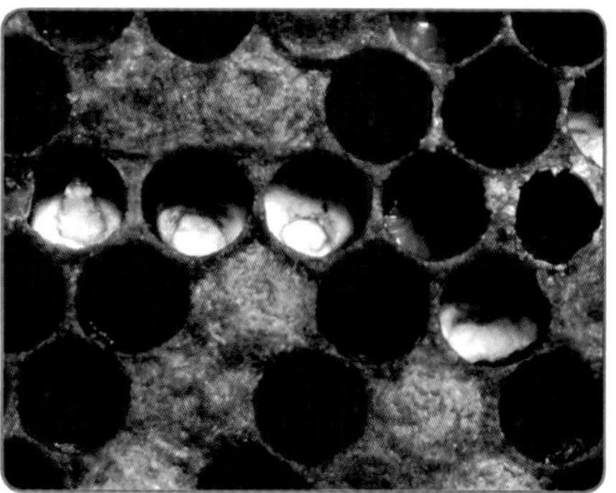

National Bee Unit, Fera (crown copyright)

Above and left: Chalk brood 'mummies' in the comb.

National Bee Unit, Fera (crown copyright)

Left: Chalk brood 'mummies' on the hive floor.

CONTROL

Some bees are more susceptible to chalkbrood than others. A severe infection will very often be reduced or eliminated by re-queening the colony from one that is less susceptible. Because the spores remain on the comb you'll reduce the incidence of the disease if you replace old brood combs on a regular basis. You don't need to replace them all at once, but changing two or three of the oldest ones each year will be beneficial.

Sac brood

Sac brood is caused by a virus which prevents the larva from making its final moult into an adult. The dead larva lies stretched out in the cell in a fluid-filled sac that can often have the appearance of a Chinese slipper with its curled up toe. The larva turns yellow and then black, finally drying into a brown scale which can be removed easily.

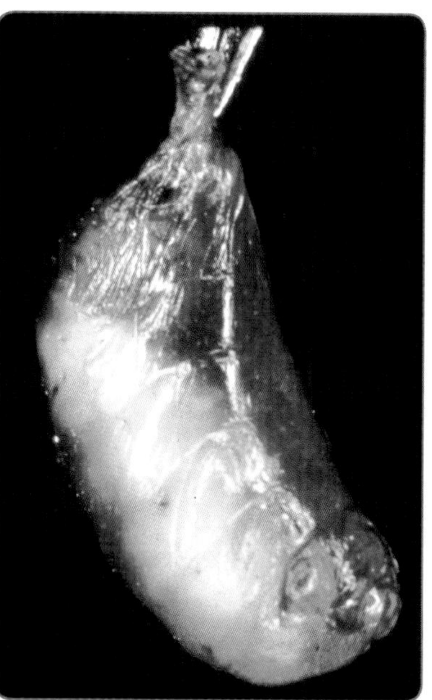

National Bee Unit, Fera (crown copyright)

Above: A larva dead from sac brood.

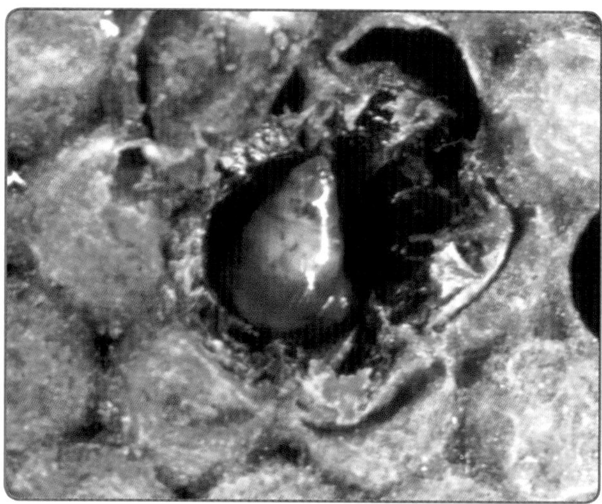

National Bee Unit, Fera (crown copyright)

Above: The typical 'Chinese slipper' effect seen with sac brood.

Sac brood rarely affects large numbers of larvae and it generally clears up without treatment. However, severe cases can be treated by re-queening from a colony free of the disease.

Bald brood

Here, the affected cells are not capped. This is often because of wax moth activity but it can be due to a genetic fault. The exposed pupae continue to develop normally and emerge as adults. Maintaining strong colonies helps the bees deter wax moth.

Below: Bald brood with uncapped pupae.

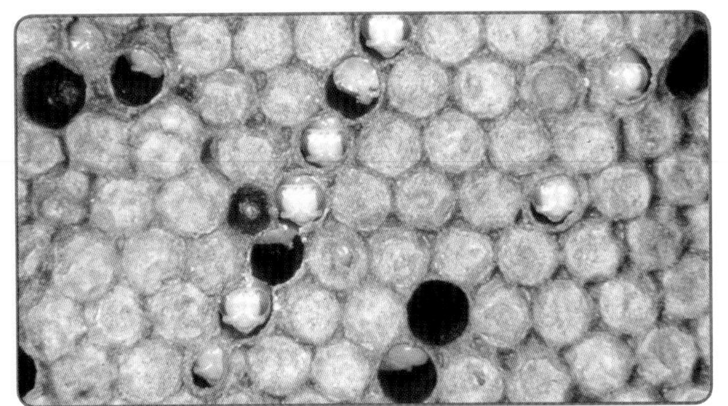

Viruses

Bees have their own viruses, which are similar to human ones in that they can be present in a colony in what is known as an inapparent infection which has no effect on the colony. It's like the human cold virus that's everywhere but only becomes active in certain circumstances.

Until the arrival in England in 1992 of the varroa mite, *Varroa destructor*, most beekeepers took little notice of bee viruses because they rarely caused a problem. However, the varroa mite feeds on the bee's blood (haemolymph) by puncturing the larval skin. This can trigger and spread viruses throughout the colony.

The commonest virus is deformed wing virus where affected adults have stunted or shrivelled wings. They're unable to fly and cannot join the foraging force, seriously weakening the colony. Without treatment to reduce the mite level, the colony will eventually die. Even if the beekeeper takes action this may not be in time to save the bees.

The main way to reduce the chance of a virus becoming active in your colonies is to keep the varroa mite level low.

Left: A worker affected by deformed wing virus.

Laying workers

Although not strictly a disease, laying workers affect colony development. There should be only one egg in each cell. Sometimes a new queen that has only just started to lay may deposit more than one, but if you find multiple eggs in lots of cells you need to determine that she's still present.

Left: Multiple eggs in a cell (top right) can indicate laying workers.

If the queen disappears for any reason, the bees can rear new queens from worker larvae that are less than three days old. However, if no such larvae are present the colony becomes what beekeepers call 'hopelessly queenless'. In other words, it cannot replace the lost queen and is doomed.

Being female, worker bees have ovaries, but pheromones produced by the queen normally inhibit them from laying eggs. Once the queen has gone, however, that restraint is removed and a few of the workers will start to lay. Because they cannot mate they can only produce drones, signalling the end for the colony. Laying workers often lay multiple eggs in a cell but they don't lay in the consistent pattern of the queen. They also tend to lay in worker comb, and the cells become distorted as they expand to accommodate the larger drone.

The presence of laying workers can be identified by:

- Multiple eggs in random cells.
- Uneven comb.
- Domed cappings on worker cells.

Below: Uneven drone cappings indicating laying workers.

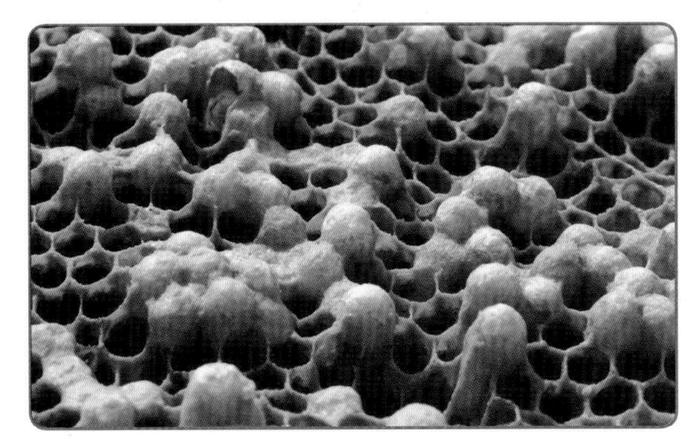

THE SOLUTION

It's very difficult to re-queen a colony with laying workers, especially if they're well established. It's best to regard the colony as lost, but the bees don't have to be destroyed. Take each frame in turn and shake the bees off on to the ground in front of another strong colony. Then remove the hive of the laying worker colony from its stand. The bees will eventually join the other colony. They'll then receive queen pheromone through trophallaxis, their urge to lay eggs will be suppressed and they'll return to normal.

The laying worker brood combs can be swapped with food combs in other colonies where they'll develop normally. Store spare food combs in a bee-tight box and use them for winter feed.

If you really want to try to keep your laying worker colony, you must persuade it to raise another queen. First try to reduce the urge of the workers to lay – brood also produces a pheromone which suppresses laying workers, so introducing a comb of mostly unsealed brood can have a restorative effect. By making sure this comb contains eggs or very young larvae, you also give the colony the material from which to raise a new queen. The bees may not do this on the first comb you introduce and you may have to repeat the procedure. The laying workers may or may not raise a queen but you may feel it's worth a try.

If you want to maintain the same number of colonies in your apiary I think it would be easier to shake out the laying workers and then take positive steps to divide a strong colony that you like in order to replace it.

Drone-laying queen

There is a slight complication here because a queen can sometimes run out of sperm. She 'measures' the width of a cell with her front legs and then fertilises the egg if it's a worker cell, or not for a drone cell. However, if she has used all the sperm stored in her spermatheca she'll end up laying unfertilised eggs in worker cells. This will mimic the laying worker situation, with drones being produced and the contours of the comb being disrupted.

If you have drones developing in worker cells you need to check carefully whether or not the queen is still there.

Below: Drone and worker capped brood indicating a drone-laying queen.

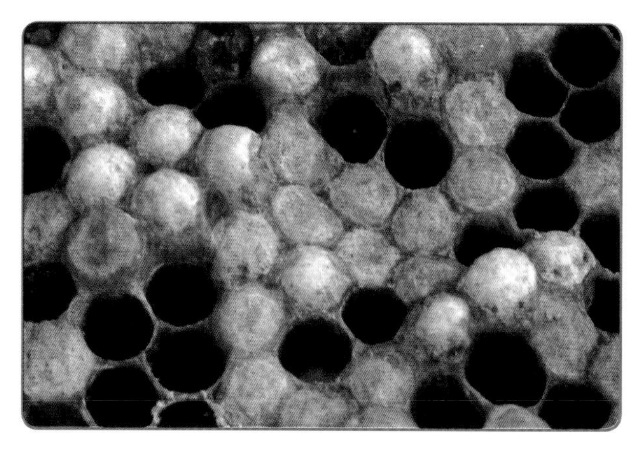

THE SOLUTION

As the colony is queenright, even if the queen can only lay drone eggs, you can replace her and introduce one that is laying normally. Queen introduction has been discussed earlier (see page 100).

Adult bee diseases

Adult bees have their own diseases. These are Nosema and Amoeba, which affect the gut; and Acarine, which affects the tracheae or breathing tubes.

Nosema

Nosema is the most significant universal bee disease. Until recently the only causative agent known was a specialised parasitic microsporidian fungus called *Nosema apis*, but recently bees have been found to be infected by *Nosema ceranae*, which originated in the Asian honey bee, *Apis cerana*. This is proving to be a more aggressive form of the disease.

The house bee cleans up mess in the hive by licking and biting. In doing so, she accidentally ingests the Nosema spores. These multiply in her gut, severely interfering with her ability to digest food, particularly pollen. This can cause dysentery, although not all bees with dysentery have Nosema. Dysentery is indicated by brown faecal streaks on the inner cover, the comb and other surfaces.

Below: Excreta on the inner cover could indicate Nosema.

Nosema also shortens the bee's life. The effect is that the colony either builds up slowly in the spring, stays the same size or even dwindles, depending on the severity of the infection. In severe cases the colony may dwindle and die. If you have more than one colony you can compare colony development rates.

IDENTIFICATION
Nosema can only be positively identified using a compound microscope, although you may have strong suspicions if your colonies don't build up in the spring. Your local association may well have a microscopist who can do this examination for you.

TREATMENT
The antibiotic fumagillin, previously available as Fumidil® B, will kill the active stage of Nosema in the bee. However, the Veterinary Medicines Directorate withdrew the UK marketing authorisation with effect from 31 December 2011 and suppliers were allowed to sell stocks only until 30 June 2011. This means there is no authorised treatment for Nosema in the UK.

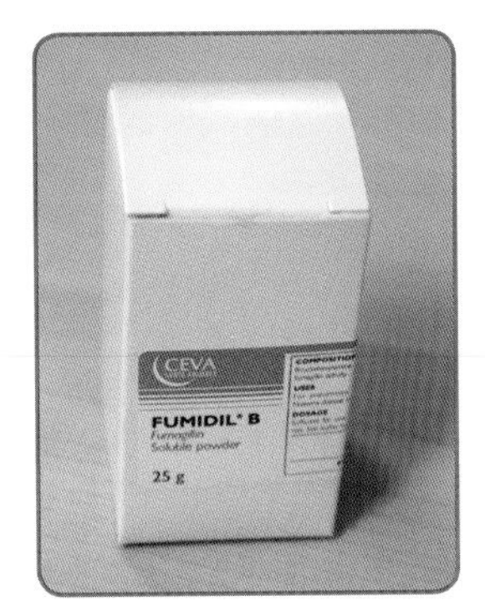

Above: Fumidil® B.

However, a break in the course of the disease and a considerable help to a return to health can be achieved by transferring the bees onto 'clean' (uninfected) comb and/or foundation. With larger colonies, move the queen on a comb of brood into a box containing clean comb and place it, over a queen excluder, on top of the affected colony. As soon as new brood is found in the upper box, move the original comb (minus the queen!) back to the lower box. As soon as all the brood has hatched in the lower box, remove it and its combs, leaving the bees on 'clean' equipment and combs.

Small, sick colonies can be united and the 'combination' colony with your preferred queen can be transferred onto clean comb as above.

Amoeba
Amoeba is caused by an organism called *Malpighamoeba mellificae* which affects the Malpighian tubules, the bee's equivalent to our kidneys. It's passed round the colony in the same way as Nosema. Amoeba doesn't appear to be a serious problem and an affected colony can be transferred on to clean comb.

Amoeba often occurs with Nosema. It can only be identified by microscopic examination as it has no outward symptoms.

Acarine or tracheal mite
Acarine is an infestation caused by a mite, *Acarapis woodi*, rather than a disease. It's also known as Isle of Wight disease, having been first identified there. The mite is also called the tracheal mite, because it invades the bee's breathing tubes or tracheae and breeds there. This shortens the bee's life, which slows colony development. A severe infestation can kill the colony.

There is currently no legal treatment for Acarine in the UK but it's not often detected today. Some imported bees appear to be more susceptible than our native bees. With no available treatment, it's probably better to let susceptible colonies die and replace them with ones that show resistance. Re-queening is also an option.

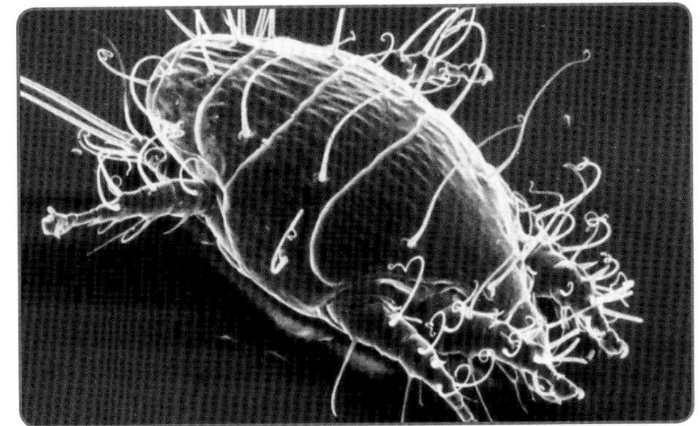
National Bee Unit, Fera (crown copyright)

Above: The Acarine mite.

IDENTIFICATION
Acarine mites can be seen under a low-powered dissecting microscope. The dead bee is pinned and the head removed to expose the first pair of trachea. For a more detailed examination, the 'collar' at the front of the thorax is also removed.

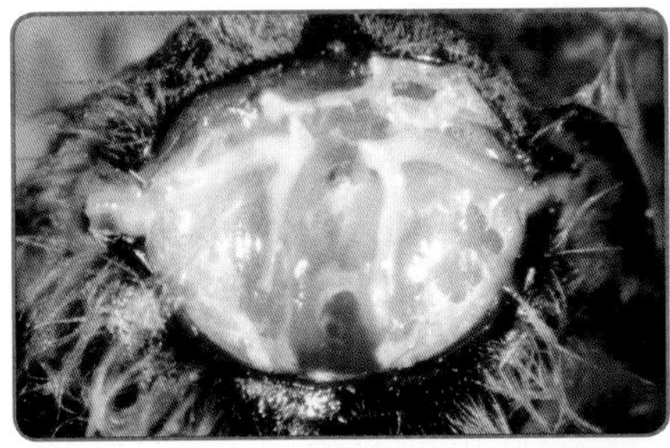
National Bee Unit, Fera (crown copyright)

Left: Dissecting a bee to inspect the trachea for Acarine mites. This one is healthy.

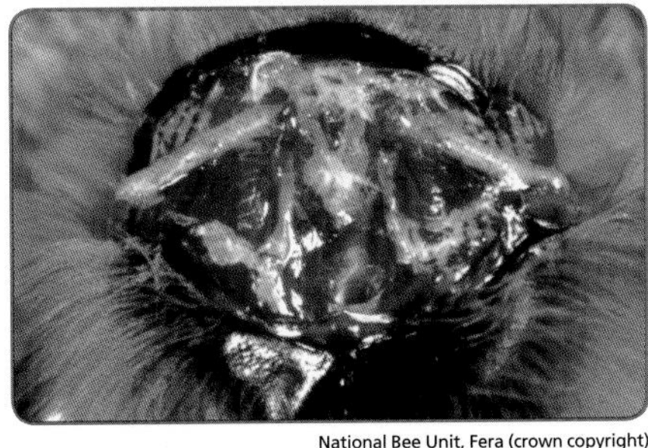
National Bee Unit, Fera (crown copyright)

Left: This bee has an Acarine mite infestation.

Varroa
The mite *Varroa destructor* is, without doubt, the most serious threat to our bee colonies. It was first described as a parasite of the Eastern honey bee, *Apis cerana*, but transferred to our Western honey bee, *Apis mellifera*, when colonies were taken to regions such as Siberia and then brought back to Europe, with their mites. In one sense it doesn't matter how varroa reached the UK. It's here and we have to deal with it.

THE VARROA LIFE CYCLE

Both adults and young feed on the bee's blood or haemolymph. Adults can survive over winter either on adult bees or in the brood, if present. They can only reproduce in the brood and prefer drone cells, which are sealed for longer than worker cells, allowing two mated daughters to emerge rather than one from worker brood. When drone brood is present, the varroa population can increase very rapidly.

IDENTIFICATION

Varroa mites are about 1.6mm (1/20in) across and can be seen with the naked eye, although on adult bees they hide between the abdominal plates. If you uncap drone brood with an uncapping fork and pull out the developing pupae, the mites are easily seen against the white bodies.

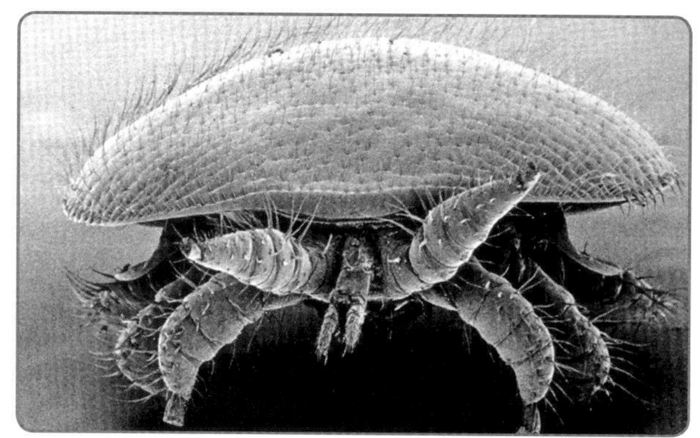

National Bee Unit, Fera (crown copyright)

Above: The varroa mite.

- Select an area where the drone pupae are at an advanced (pink-eyed) stage.
- Slide the tines of a honey uncapping fork under the cappings.
- Lift out the pupae, count and record their number.
- Estimate and record the number of mites present.
- Repeat until you have checked 100 pupae.
- Divide the total number of mites by the total number of pupae.
- A 5% infestation is light; over 25% is severe.

Below: Uncapping drone brood to check for varroa mites.

National Bee Unit, Fera (crown copyright)

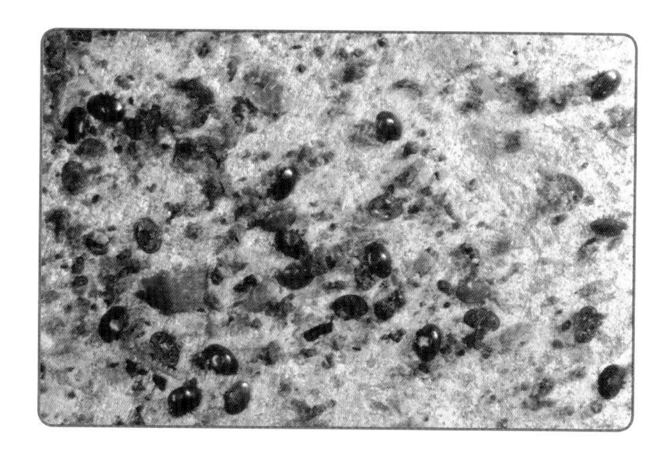

Above: Varroa mites on the open-mesh floor insert.

With an open-mesh floor you can monitor mite levels. The floor includes a mesh panel through which mites can fall but bees can't pass. Fallen mites cannot climb back into the hive. The debris, which also includes pieces of wax, bee parts, pollen and dirt, falls on to a removable tray which you can inspect to count the number of dead mites.

- Examine the floor debris and record the number of mites.
- Divide this by the number of days since your last count to get an average daily mite drop.
- The average daily mite drop depends on the varroa population in the colony and the amount of emerging brood.
- Check whether your colony needs urgent treatment.

A colony with an average daily mite drop of 0.5 in winter/spring is likely to collapse by the end of the season without treatment. Similarly, the colony needs treatment if you have an average daily mite drop of 6 in May, 10 in June, 16 in July, 33 in August or 20 in September.

TREATMENT

The original chemical varroa treatment was synthetic pyrethroids. Impregnated plastic strips, available as Bayvarol® and Apistan®, were hung between the frames in the brood nest for six weeks when there was no honey on the hive. However, the mites adapted and became resistant to the chemical so other treatments had to be developed.

Left: Inserting an Apistan® strip.

Bayvarol® and Apistan® are still available and they have been joined in the UK by Apiguard®, ApiLife Var® and Thymovar® which are all based on thymol and other essential oils. Hive cleansers are also available under the trade names Varroa-Gard®, Exomite® and BeeVital® Hive Clean. Oxalic acid and formic acid are also used as treatments although they are not legally permitted in the UK. Applicators for accurate dosing and minimal risk to the beekeeper are available.

Whichever treatment you use, follow the instructions carefully. Don't be tempted to think that if you apply twice as much and/or leave it in the hive for twice as long the treatment will be twice as effective. What you'll be doing unwittingly is building mite resistance to the product, which will then become useless.

INTEGRATED PEST MANAGEMENT

Integrated pest management (IPM) is a system of attacking varroa mites in various ways to keep populations low. This is the only way you have of controlling viruses, as high mite numbers appears to be one of the triggers that activates them. IPM does not exclude chemical treatments but uses them when required in conjunction with other biotechnical and management techniques.

The fact that varroa mites favour drone brood can be used to our advantage. Removal of drone brood does not have a significant effect on a colony unless the drones are needed for a breeding programme. Between mid-March and the end of May, take a shallow frame, preferably with drawn comb, and insert it next to the outside frame containing brood. The bees will build drone comb below the bottom bars and, when this is sealed, complete with varroa mites, you can remove it and destroy it.

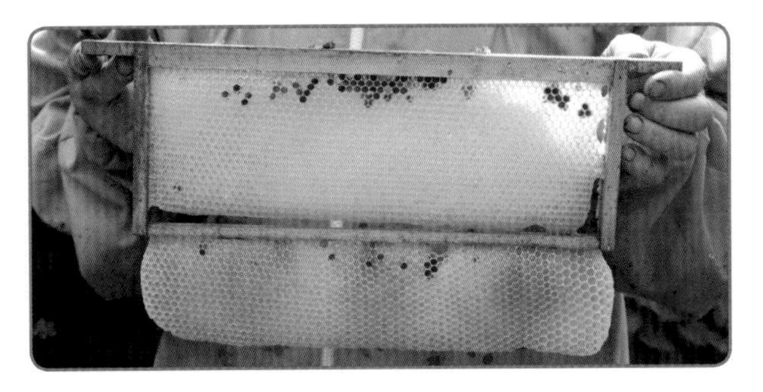

Above: Sacrificial drone brood built on the bottom of a super frame.

What you *must* do with this method is to remove the sealed brood *before* the drones emerge. If you don't, you'll be producing mites in your colony very efficiently.

Varroa mites can also be removed from the colony by confining the queen on successive brood combs so that this is the only brood in the colony and hence the only place for varroa to breed. Again, you must remove the frame and destroy the comb (and mites) before the brood emerges.

Mite populations can be reduced by modifying the shook swarm. Destroy all the combs that are removed from the colony. Include two or three drawn combs with the foundation in the new brood box, and the queen will lay in these. The brood will attract mites transferred with the adult bees. When it's sealed, the frames can be replaced with foundation, placed on the outside of the colony, and brood and mites destroyed.

Tropilaelaps mites

Tropilaelaps mites are primarily parasites of the giant Asian honey bee, *Apis dorsata*, although, just like varroa, they've transferred to *Apis mellifera*. *Tropilaelaps* is thought to be restricted to tropical or sub-tropical regions but it could survive in the UK and has been declared a notifiable disease here. If you suspect its presence, you have a legal obligation to inform the authorities.

The mites are much smaller than varroa, at around 1mm long and 0.6mm wide. They're a light reddish brown. They reproduce in sealed brood cells, particularly drone, but unlike Varroa cannot feed on adult bees. They're detected by uncapping drone brood and inspecting the pupae.

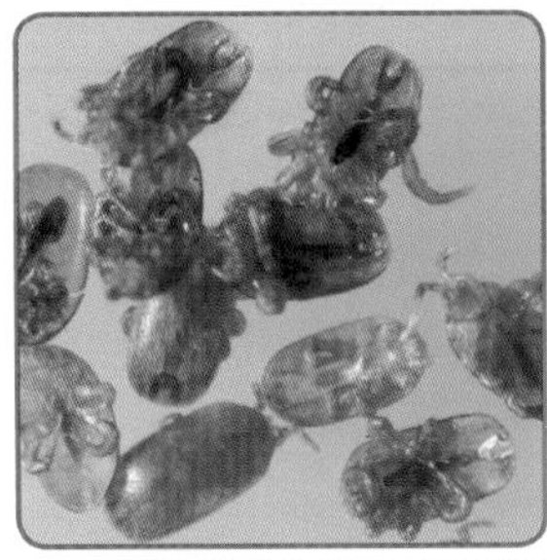

National Bee Unit, Fera (crown copyright)

Above: Small *Tropilaelaps* mites.

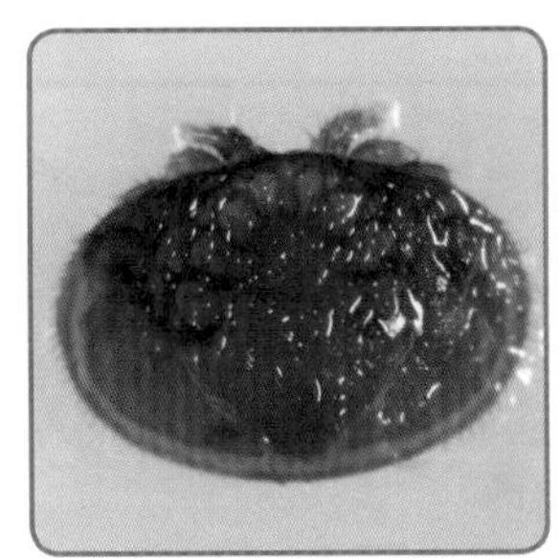

National Bee Unit, Fera (crown copyright)

Above: The larger varroa mite.

Small hive beetle

Small hive beetle (*Aethena tumida*) originated in South Africa but has now found its way to Australia and the USA. The black beetles, which are 5–7mm long, can be identified by their clubbed antennae, and the beige larvae, which are 10–11mm long, by spines on their backs and three pairs of legs behind the head.

National Bee Unit, Fera (crown copyright)

Left: An adult small hive beetle.

National Bee Unit, Fera (crown copyright)

Left: The typical clubbed antenna of the adult beetle.

The female beetle enters the colony and lays eggs in nooks and crannies. The larvae prefer to eat bee eggs and brood rather than honey and pollen. They spoil the combs by leaving slime all over them. To pupate, the larvae exit the hive and burrow into the surrounding soil, the emerging adults entering the hive to repeat the cycle.

Small hive beetle causes severe economic damage and it has been declared a notifiable disease in the UK. It has not yet been found here but every beekeeper must keep a close check so that, should it arrive, the authorities are notified as soon as possible.

National Bee Unit, Fera (crown copyright)

National Bee Unit, Fera (crown copyright)

Above: Small hive beetle larvae.

Left: A pupating larva.

IDENTIFICATION

- When you open a colony, look for beetles running away from the light.
- Remove the supers (and upper brood chamber of a double-brood colony) and stand them on an upturned roof. Cover with an inner cover.
- After a few minutes, lift the boxes and look for beetles in the roof.
- Examine all the brood combs carefully and check the floor for larvae.
- Eggs are about two-thirds the size of bee eggs. Look for clusters in cracks and crevices.
- Use a beetle trap to check whether they're present. These mostly work by providing a dark space where the beetles can hide.

Colony Collapse Disorder

There has been much media coverage of Colony Collapse Disorder (CCD), first reported in the USA. While there was a greater than usual loss of colonies reported in the UK in 2008 and 2009, I personally do not believe that CCD is present here and that these losses are more likely to be a result of varroa and poor weather conditions.

SYMPTOMS

- The majority of the adult bee population leaves the colony over a short period (hours or days).
- These bees are not found in the vicinity of the hive as would be the case if they had been killed by something toxic.
- The queen remains in the hive with a small number of workers.
- A colony that has died from CCD is not immediately robbed out by bees from other colonies as would normally be the case.
- There is an unusual delay before the dead colony is invaded by wax moths.

A great deal of money is being invested in research into the causes of CCD. There are several theories, but what is clear is that it does not have a single cause. At the time of writing the latest theory is that it's a combination of *Nosema ceranae* and a newly discovered insect iridescent virus. Hopefully the researchers will come up with an answer, and a solution, soon.

Robbing & drifting

These are not diseases but are a major method of their spreading, the other being beekeepers transferring combs from a diseased colony to a healthy one.

Below: Wasps squeezing into a hive to rob honey.

Robbing

Bees from one colony will try to illegally enter and steal honey from another. If the colony being robbed is diseased or has died from AFB, EFB, Nosema or anything else, the robbing bees will pick this up and carry it back to their own colony.

It's much easier to prevent the advent of robbing than to stop it once it has started. Make sure that there are no gaps between your hive boxes where bees (or wasps) can gain access. At 6–8mm (¼–⅓in), the bee space is really a very small gap. Block any gaps with small pieces of foam, such as from cushion squabs. Plasticine can be moulded to plug the holes. Alternatively, tape over them with something like duct tape, although you'll need to replace this if you separate the boxes for any reason.

These measures must be regarded as temporary and you should note which boxes need to be repaired once they're away from the bees. Keeping your equipment in good order will pay dividends. I once had the honey from four full supers in an out-apiary stolen by wasps because I didn't notice a couple of small gaps between them.

Drifting

Diseases are also spread by drifting. Bees returning from a foraging flight can accidentally enter the wrong hive. Guard bees will accept strangers carrying a crop full of nectar and let them in. However, this is also a prime way to share diseases. The prevailing wind tends to carry bees along a row of hives and the population in the end ones increases at the expense of those in the middle. These colonies will 'collect' more honey and will have more supers. Bees seem to be attracted to taller hives and this just exacerbates the situation.

Although you might like a tidy apiary with a formal hive arrangement, returning bees find repeated patterns confusing. We described earlier (page 45) that if you keep your hives in regular blocks you need a distinct landmark, such as a shrub or bush, to identify each one. Hive entrances should also face in different directions. If you keep two hives on the same stand, make sure that they're at least 45–60cm (1½–2ft) apart.

Below: Two hives on a stand should be at least 40 cm apart.

Apiary hygiene

Make sure you don't leave bits of comb or honey lying around in the apiary. Take a small covered container to the apiary to collect pieces of brace comb or scrapings from the top bars or excluder. Keep it covered so that bees don't find it and try to rob the contents. You can wrap up the contents and dispose of them or, if you have a solar wax extractor, you can collect them in there and recycle the wax.

The National Bee Unit

Beekeepers in England and Wales are very fortunate to have the services of the National Bee Unit (NBU), which is part of the Food and Environment Research Agency (Fera), based at Sand Hutton near York.

The NBU operates a disease inspection service. Regions of England and Wales each have a full-time Regional Bee Inspector (RBI), supported by a team of Seasonal Bee Inspectors (SBIs) who will come and check your colonies for the notifiable bee diseases: AFB, EFB, small hive beetle (SHB) and *Tropilaelaps* mites.

If a notifiable disease is confirmed, the appropriate treatment will be given. The bee inspector will issue you with a Standstill Notice, prohibiting the removal of bees and equipment from your apiary. If there are no further signs of disease in six to eight weeks, the Standstill Notice will be lifted. The SBI will contact all known beekeepers within a 5km (three-mile) radius and check their colonies for the disease.

The NBU produces excellent descriptive leaflets on bee diseases which can be downloaded from its website, www.nationalbeeunit.com. This website also contains much useful information on diseases. If you live in England or Wales, I very strongly advise you to register on BeeBase. By knowing where beekeepers and their colonies are, the NBU is best placed to help all of us fight disease and keep our colonies healthy. Recently, information for the Scottish disease inspection service has been incorporated into BeeBase.

Pests

As well as diseases our honey bees also have to contend with a number of pests.

Mice

Mice can get into colonies and into stored equipment. They'll generally only be able to enter occupied hives when the bees aren't active, that is when they're clustering for winter. Once inside, mice can cause havoc. They'll chew wooden parts, eat stores, chew holes in comb and bring in nesting materials.

I understand mice can enter a gap as small as 12mm (½in), which means they cannot get through one only a bee-space (6–8mm) high. Thus a shallow floor will keep mice out. You'll have to protect a deep entrance with a mouseguard. This is a metal strip that's wide enough to easily fit the width of the hive and cover the entrance. It has 10mm (⅜in) holes punched in it that allow bees through but nothing larger.

Take the entrance block out and fit your mouseguards by late October/early November. With stacks of equipment, a spare queen excluder at the top and bottom will keep mice out.

Above: A mouse which gained access to a hive but didn't survive.

Rats

Rats can cause even more damage than mice. They pose an additional problem if they get into stored equipment because as they move about they dribble urine, which can carry Weil's disease. If they get into your equipment, replace combs with new foundation and scrub other hive parts thoroughly with a strong disinfectant before using them again.

Below: Using a mouseguard to keep out rodents in winter.

Wasps

Wasps can be a real pain to bees when they try to rob honey stores from hives in late summer, but they're only looking for food. Adults feed on sugar and will work flowers for nectar. They feed their larvae on protein, such as flies and bees, and are rewarded with a sugary secretion. The wasp colony has an annual life cycle, with new, mated queens hibernating over the winter. As the colony declines in the autumn there are fewer larvae to provide sugary rewards and the adults look elsewhere, including in beehives.

If your hive is bee-tight it will also be wasp-tight, and this is the best way to help your bees defend their stores. As well as making sure all boxes and roofs fit snugly, check that the ventilation holes in the roof are covered with a fine gauze. Next reduce the size of the hive entrance to one that the colony can defend easily. Smaller colonies require smaller entrances, even down to a single bee-space. It's better for bees to queue up to go in and out than that wasps sneak in round the corner. Moving the brood nest to just behind the entrance will also encourage the colony to defend it more strongly as the guard bees are positioned closer to the point of attack. Consider uniting a colony too small to help in this way with a larger one (see page 96). Kill the queen in the colony you like least and put the weakest of the two on top of the newspaper.

Below: A colony needs to be able to defend its entrance against intruders, especially in autumn.

Above: Bees queuing to enter the hive.

Bees

Given the opportunity, bees will rob other colonies. It's far better to prevent this than to try to stop it. Again, make sure the hive is bee-tight and reduce the entrance. You should be aware of potential robbing when you're examining your colonies. Don't do so during the main part of the day if possible, but if you have to, close the hive as soon as you get a hint that robbing is starting.

Above: Woodpecker damage.

Birds

Birds eat insects, including bees. However, the numbers lost are generally very small and colonies can cope and replace them.

A bird that can cause considerable problems is the green woodpecker. A beehive is like a hollow tree with a large number of insects inside, which is a feast worth having. Woodpeckers can ignore beehives for years and then suddenly 'discover' them. After that they'll return year after year. They can smash large holes in the sides of boxes and eat bees, wax and honey. Generally they home in on occupied hives but I have known them damage even stored equipment.

The best protection is to wrap hives in small-mesh (25mm/1in) chicken wire. Make sure it's deep enough to cover the whole depth of the box to the floor and also fold over the roof to keep it in place. Keep the wire over the entrance as the bees can fly through and woodpeckers cannot damage the entrance slot. Don't remove the mesh until March/April. Roll it up carefully and it will last for years.

A less expensive alternative is to pin strips of heavy-duty plastic to the roof, making sure that these hang down to the floor. You can hang wide panels down each side or cut thinner strips that will flap in the wind and hopefully provide an additional deterrent. Cut the strips slightly shorter over the entrance so that the bees can still fly in and out.

Livestock

If your apiary is likely to be invaded by neighbouring livestock, you need to fence it. Don't take the risk of your hives being knocked over. If this happens in winter the colony could die.

Above: Using wire netting to protect against woodpeckers.

Vandals

Unfortunately, some humans find pleasure in messing with beehives. They throw stones at hives and knock them over. One beekeeper told me that vandals had put lighted banger fireworks into his hives. Why people do this sort of thing I will never understand, but the best way of preventing it is to keep your hives out of sight. Alternatively you can paint them in subdued camouflage colours to disguise them.

Below: Flapping strips of plastic also deter woodpeckers.

Wax moths

These come in two varieties: the greater wax moth (*Galleria mellonella*) and the lesser wax month (*Achroia grisella*). Both will eat beeswax and can be a serious pest, especially of stored comb. The greater wax moth prefers brood comb and the lesser variety is more a pest of stored supers. When the greater wax moth pupates, it chews grooves into the wooden hive surfaces. Wax moths can ruin combs and produce copious amounts of webbing and frass. Avoid wax moth damage by keeping your colonies strong so that the bees can keep an infestation at bay. Don't store your supers in a warm place, which encourages the moths to breed.

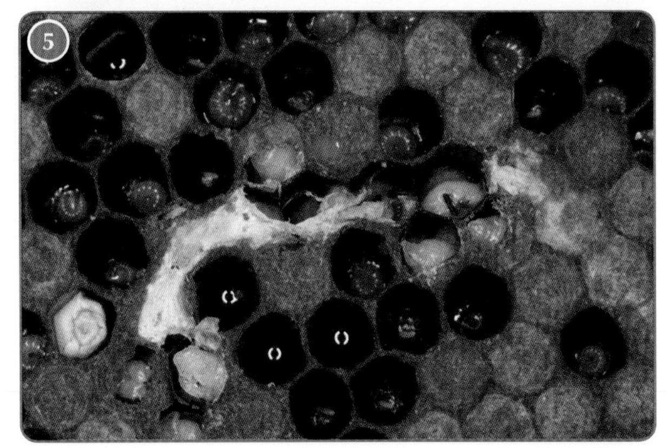

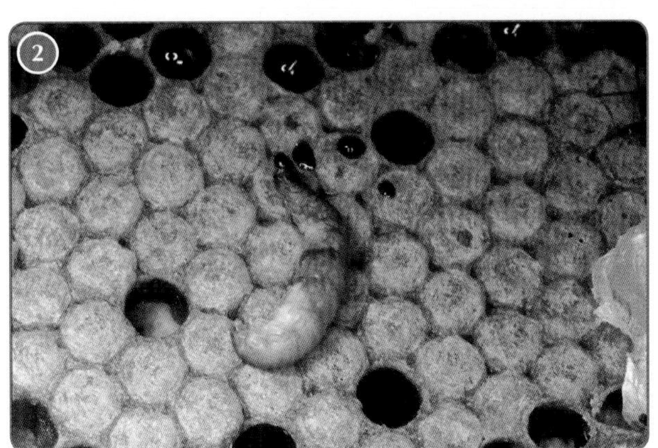

Greater wax moth:
1 Adult moth (dead)
2 Larva
3 Grooves in a brood box where larvae have pupated

Lesser wax moth:
4 Adult moth
5 Trail left by a larva burrowing through the comb
6 Webbing and frass

There are several storage alternatives you can consider:

- Store piles of empty supers on a stand outside. Put mouse-proof mesh (such as a queen excluder) at the top and bottom of the pile and cover with a watertight roof.
- After you've prepared a colony for winter, store supers over the inner cover with the feed hole open and the roof on top.
- Wrap combs in plastic and store them in a deep freeze or cold place that's inaccessible to mice.
- Treat stored supers with *Bacillus thuringiensis*, a biological control which kills young wax moth larvae.

THE BEEKEEPING YEAR

The beekeeping year

This chapter is designed to give you a quick summary of what your bees will be doing and how you should be looking after them through the year. Bear in mind that the actual timings will vary according to where you live. Take them as a guide and apply your local knowledge and common sense. Get into the habit of keeping records and then you can refer back to these to check when things happen in your part of the country.

January

- Visit your apiary occasionally to check that all is well, that no hives have been turned over and no mouseguards have become detached.
- Bees should be flying on a sunny day when it's warm enough. They'll defecate and may be able to collect pollen from winter-flowering plants.
- If you're likely to have snow in your area, shade the entrance to discourage the bees from flying. They can be confused by reflection off the snow, fly into it and so perish.

Below: Colonies in winter.

Above: Hefting a hive to check a colony has sufficient food.

February

- Increasing day length means that the queen is starting to lay.
- Greater activity means that bees will start making inroads into their winter stores.
- Heft the hive to check that the colony has sufficient food.
- Crumbs of beeswax on the floor or alighting board indicate bees are opening the cells and using their stores.
- Pollen loads being taken into the hive indicate increased brood-rearing activity.
- Make sure that the water source near the apiary is in a sunny spot that will warm the water before the bees drink it. A chilled bee will be unable to fly back to the colony.

March

- Make sure all your colonies are flying well when weather conditions are suitable.
- Check for large pollen loads being taken into the hive.
- On a warm day, use a little smoke and lift the inner cover. The bees should look as strong now as they did last September.
- Look for visible excreta on the top bars, which is a sign of dysentery.
- Remove the props under the inner cover and close the feed hole.
- Heft the hive and, if necessary, feed dilute syrup (1:1 sugar:water) in a contact feeder over the winter cluster.
- Don't feed until the weather is good enough for the bees to take flights to defecate on a nominal weekly basis.
- Don't look at the brood nest too early or the bees may kill the queen. Wait for a warm day. When flowering currant (*Ribes sanguinium*) is in flower, you should be able to inspect your bees on a warm day.

Above: Pollen loads of spring.

April

- In mid-April, start regular brood nest inspections of larger colonies.
- Keep records of what you see.
- Look out for signs of swarming.
- Add a queen excluder and super when the brood box is full of *bees*.
- Remove any old, broodless combs and replace them with frames of foundation.
- Feed the colony if necessary to help them draw out the comb.
- Ask an experienced beekeeper for help if you think things aren't quite right.

Below: Regular brood inspections.

Above: Queen cells when a colony decides to swarm.

May

- Make sure that colonies have sufficient room.
- Continue adding supers as each new box on the hive fills with *bees*.
- Continue regular brood nest inspections.
- List the steps of your chosen method of swarm control.
- Make sure that you have all the necessary equipment to hand.
- Remove supers or frames full of honey that are ready for extraction.
- Replace with empty supers to prevent colony congestion.

June

- Continue as in May.
- If you've successfully controlled swarming, the need for very regular inspections is coming to an end.
- If you have a 'June gap' in your district, make sure that your bees have sufficient food reserves.
- Try to make sure that a newly mated queen resulting from swarm control measures is heading your main honey-producing colony. This will greatly reduce the chance of the colony swarming a second time.

Below: A newly mated queen.

Left: Bees pollinating a field of borage.

July

- July is still the main honey month for most beekeepers, with the lime and clover coming into flower.
- Estimate how much honey you have in your supers.
- When the bees seem to have stopped storing honey freely in the supers, remove most of your honey crop when the cells are sealed.
- Don't add more supers unless you know that your bees need them. At this time of year the brood nest is beginning to contract.
- Be aware that your bees could still be collecting nectar from crops like borage.

August

- Remove the supers and extract them.
- Feed your colonies if they have no food reserves in the brood box.
- Don't leave frames or honey around in the open.
- Make hives bee-tight and reduce entrances to prevent robbing.
- Don't open a colony unless absolutely necessary.
- Start varroa treatment if you're using Apiguard®, ApiLife Var®, Apistan®, Bayvarol® or Thymovar®.

Below: Applying Apiguard® to control varroa.

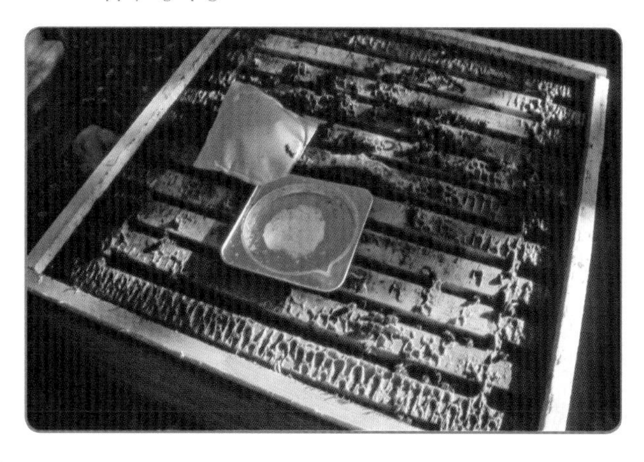

September

- Assess the amount of honey in the brood nest.
- Feed the bulk of the winter stores required.
- Aim to get stores completely surrounding the brood nest.
- Heft the hives to check their weight.
- Ivy will be coming into flower but don't rely on this to provide winter stores, as bad weather may not allow your bees to forage on it.
- Store supers to reduce the chance of damage by wax moth. If necessary, treat with *Bacillus thuringiensis* available as B104 (Certan) or Mellonex.

Below: Feeding with sugar syrup in the evening.

Above: Reducing the entrance to prevent robbing.

October

- Check whether your colonies are being bothered by wasps or robber bees.
- Make sure all hives are bee-tight and wasp-tight.
- Keep entrances small until the first frosts have killed the wasp nests.
- Remove the entrance block and fit a mouseguard to deep entrances.
- Remove varroa treatments according to the manufacturer's instructions.
- Record the batch numbers of treatments used and the dates of insertion and removal. This is a legal requirement.

November

- Raise inner covers for winter ventilation.
- Protect hives against woodpeckers if they're a problem in your area.
- Protect hives against livestock.
- In windy locations, place a weight on the roof to prevent it blowing away.
- Leave your bees alone. You must not disturb the winter cluster.

Below: Plan to attend apiary meetings and training sessions.

Above: Woodpecker protection.

December

- Check on your bees every couple of weeks to make sure all is well.
- Use your records to review the past year.
- Identify what worked and why.
- Determine what didn't work and why not.
- Plan your activities for the coming year – colony increase, queen rearing, migratory beekeeping.
- Make or buy any additional equipment you'll need. If you can saw straight, you can make a beehive!

Below: Golden mead.

Further reading

There are many useful books on bees, beekeeping and associated subjects. I have tried to list a few dealing with each aspect of beekeeping, but for a wider selection contact the bee booksellers and appliance manufacturers. Some of the books listed are out of print but it's worth trying to obtain them. Your local beekeeping association may well have copies in its library or you may be able to borrow them from your public library. Alternatively you may be able to buy them second-hand on the Internet – www.bookfinder.com is a good place to start.

Andrews, S.W. *All About Mead* (Northern Bee Books, Mytholmroyd, 1982).

Aslin, Pauline. *How to Make Beeswax Skin Creams for Use at Home* (Bee Craft Ltd, Stoneleigh, 2010).

Aston, David, and Bucknall, Sally. *Plants and Honey Bees: Their Relationships* (Northern Bee Books, Mytholmroyd, 2004).

—— *Keeping Healthy Honey Bees* (Northern Bee Books, Mytholmroyd, 2010).

Badger, Michael. *How to Use a Horsley Board for Swarm Control* (Bee Craft Ltd, Stoneleigh, 2009).

Bailey, L., and Ball, B.V. *Honey Bee Pathology, 2nd edition* (Academic Press, London, 1991).

Bee Craft Ltd. *The Bee Craft Apiary Guides to Bee Diseases* (Bee Craft Ltd, Stoneleigh, 2005).

—— *The Bee Craft Apiary Guides to Integrated Pest Management* (Bee Craft Ltd, Stoneleigh, 2005).

—— *The Bee Craft Apiary Guides to Record Keeping* (Bee Craft Ltd, Stoneleigh, 2007).

—— *The Bee Craft Apiary Guides to Swarming and Swarm Control* (Bee Craft Ltd, Stoneleigh, 2007).

—— *The Bee Craft Apiary Guide to Colony Make-up* (Bee Craft Ltd, Stoneleigh, 2010).

Benjamin, A., and McCallum, B. *A World Without Bees* (Guardian Books, London, 2008).

Brown, R.H. *Beeswax, 2nd edition* (Bee Books New and Old, Burrowbridge, 1989).

Collins. *Collins Beekeeper's Bible* (HarperCollins, London, 2010).

Davis, Celia F. *The Honey Bee Inside Out*, 2nd edition (Bee Craft Ltd, Stoneleigh, 2011).

—— *The Honey Bee Around and About* (Bee Craft Ltd, Stoneleigh, 2007).

Food Standards Agency website at http://www.food.gov.uk/.

Free, John B. *Pheromones of Social Bees* (Chapman and Hall, London, 1987).

Gibb, Andrew. *Setting up and Managing an Apiary Site* (Bee Craft Ltd, Stoneleigh, 2010).

Goodwin, Mark, and Taylor, Michelle. *Control of Varroa: A Guide for New Zealand Beekeepers*, revised edition (New Zealand Ministry of Agriculture and Forestry, Wellington, 2007).

Gould, James L., and Gould, Carol Grant. *The Honey Bee* (Scientific American Library, New York, 1988).

Gregory, Pam, and Waring, Claire. *Keeping Bees* (Flame Tree Publishing, London, 2011)

Hansen, Henrik. *Honey Bee Brood Diseases* (L. Launso, Copenhagen, 1978).

Hodges, Dorothy. *The Pollen Loads of the Honeybee: A Guide to their Identification by Colour and Form* (International Bee Research Association, Cardiff, 1984).

Hooper, Ted. *Guide to Bees and Honey*, updated edition (Northern Bee Books, Mytholmroyd, 2010).

Kirk, William. *A Colour Guide to the Pollen Loads of the Honey Bee*, 2nd edition (International Bee Research Association, Cardiff, 2006).

Lindauer, Martin. *Communication among Social Bees* (Harvard Books in Biology, No 2) 2nd edition (Harvard University Press, Cambridge, Maryland, 1978).

Morse, Roger A., and Flottum, Kim (eds). *Honey Bee Pests, Predators and Diseases*, 3rd edition (The AI Root Co Ltd, Medina, Ohio, 1997).

Munn, Pamela (ed). *Beeswax and Propolis for Pleasure and Profit* (International Bee Research Association, Cardiff, 1998).

Riches, Harry. *Insect Bites and Stings: A Guide to Prevention and Treatment* (International Bee Research Association, Cardiff, 2003).

—— *Mead: Making, Exhibiting and Judging* (Bee Books New and Old, Burrowbridge, 2009).

—— *Medical Aspects of Beekeeping* (Norwood, HR Books, 2000)

Schramm, Ken. *The Compleat Meadmaker* (Brewers Publications, Boulder, Colorado, 2003).

Seeley, Thomas D. *The Wisdom of the Hive: The Social Physiology of Honey Bee Colonies* (Harvard University Press, Cambridge, 1995).

—— *Honeybee Democracy* (Princeton University Press, Princeton, New Jersey, 2010).

Tautz, Jürgen. *The Buzz about Bees* (Springer-Verlag, Berlin, 2008).

Turnbull, Bill. *The Bad Beekeepers Club* (Sphere, London, 2010).

Von Frisch, Karl. *The Dance Language and Orientation of Bees* (Harvard University Press, Cambridge, Maryland, 1967).

Waring, Adrian. *Better Beginnings for Beekeepers*, 2nd edition (BIBBA, Doncaster, 2004).

—— and Waring, Claire. *Get Started in Beekeeping* (Teach Yourself) (Hodder Education, London, 2010).

Wedmore, E.B. *A Manual of Beekeeping*, 3rd revised edition (Bee Books New and Old, Burrowbridge, 1989).

White, Joyce, and Rogers, Valerie. *Honey in the Kitchen*, revised edition (Bee Books New and Old, Charlestown, 2000).

—— *More Honey in the Kitchen*, revised edition (Bee Books New and Old, Charlestown, 2001).

Williams, John. *Starting Out with Bees* (Bee Craft Ltd, Stoneleigh, 2010).

Winston, Mark L. *The Biology of the Honey Bee* (Harvard University Press, Cambridge, Maryland, 1987).

Useful contacts

BEE BOOK SUPPLIERS

C. Arden Bookseller
Radnor House
Church Street
Hay-on-Wye
Hereford HR3 5DQ
www.ardenbooks.co.uk

Bee Books New and Old
Ash View
Tump Lane
Much Birch
Hereford HR2 8HP
www.honeyshop.co.uk

International Bee Research Association (IBRA)
16 North Road
Cardiff CF10 3DY
www.ibra.org.uk

Northern Bee Books
Scout Bottom Farm
Mytholmroyd
Hebden Bridge
West Yorkshire HX7 5JS
www.beedata.com

BEE DISEASES INSURANCE

Colonies can be insured against losses resulting from European foul brood and American foul brood by
Bee Diseases Insurance Ltd
57 Marfield Close
Walmley
Sutton Coldfield
West Midlands B76 1YD

BEEKEEPING ASSOCIATIONS

England
British Beekeepers' Association
National Beekeeping Centre
Stoneleigh Park
Kenilworth
Warwickshire CV8 2LG
www.britishbee.org.uk

Scotland
The Scottish Beekeepers' Association
20 Lennox Row
Edinburgh EH5 3JW
www.scottishbeekeepers.org.uk

Wales
The Welsh Beekeepers' Association
secretary@wbka.com
www.wbka.com

Northern Ireland
The Ulster Beekeepers' Association
www.ubka.org

The Institute of Northern Ireland Beekeepers
105 Cidercourt Road
Crumlin
Co Antrim BT29 4RX
www.inibeekeepers.com

Republic of Ireland
The Federation of Irish Beekeepers' Associations
Ballinakill
Enfield
Co Meath
www.irishbeekeeping.ie

BEEKEEPING MAGAZINES

An Beachaire, published monthly by
The Federation of Irish Beekeepers' Associations
Scart
Kildorrery
Co Cork
Republic of Ireland

BBKA News, published monthly by
The British Beekeepers' Association
National Beekeeping Centre
Stoneleigh Park
Warwickshire CV8 2LZ
www.britishbee.org.uk
(Available only as part of Association membership)

Bee Craft, published monthly by
Bee Craft Ltd
107 Church Street
Werrington
Peterborough PE4 6QF
www.bee-craft.com

Bee World, published quarterly and available to members of the
International Bee Research Association (IBRA)
16 North Road
Cardiff CF10 3DY
www.ibra.org.uk

Beekeepers Quarterly,
published quarterly by
Northern Bee Books
Scout Bottom Farm
Mytholmroyd
Hebden Bridge
West Yorkshire HX7 5JS
www.beedata.com

Beekeeping, published ten times a year by Devon Beekeepers' Association
2 Abbey Gardens
Buckfast
Devon TQ12 0EU

Gwenynwyr Cymru,
published quarterly by
The Welsh Beekeepers' Association
61 Fir Court Avenue
Churchstoke
Montgomery
Powys SY15 6BA

Journal of Apicultural Research (electronic version only)
International Bee Research Association (IBRA)
16 North Road
Cardiff CF10 3DY
www.ibra.org.uk

The Scottish Beekeeper,
published monthly by
The Scottish Beekeepers' Association
Milton House
Main Street
Scotlandwell
Kinross KY13 9JA
(Available only as part of Association membership)

Most local beekeeping associations also publish their own magazine or newsletter.

BEEKEEPING EQUIPMENT MANUFACTURERS AND SUPPLIERS

Agri-Nova Technology Ltd
The Old Forge
Wendens Ambo
Saffron Walden
Essex CB11 4JL
www.agri-nova.biz

Andermatt BioVet AG
Stahlermatten 6
CH-6146 Grossdietwil
Lucerne
Switzerland
www.biovet.ch

Apimaye UK
Unit 9
Tower Square
Huntingdon
Cambridgeshire PE19 7DT
www.apimaye.co.uk

BBwear
1 Glyn Way
Threemilestone
Truro
Cornwall TR3 6DT
www.bbwear.co.uk

Bee Basic
5 Hillcrest Avenue
Pinner
Middlesex HA5 1AJ
www.beebasic.co.uk

Bee-Bay
Unit 1 & 2
Fletcher Industrial Estate
Clovelly Road
Bideford
Devon EX39 3EU
www.bee-bay.net

Bee Equipped
Brunswood Farm
Brunswood Lane
Bradley
Ashbourne
Derbyshire DE6 1PN
www.beeequipped.co.uk

Bee Hive Bits
Forest Farm Cottage Industries
Fockbury Road
Dodford
Bromsgrove B61 9AW
www.beehivebits.co.uk

Bee Proof Suits
Bowns Hill
Matlock
Derbyshire DE4 5DG
www.beeproofsuits.com

BJ Sherriff International
South Cornwall Honey Farm
Carclew
Mylor
Falmouth
Cornwall TR11 5UN
www.beesuits.com

Blue Bell Hill Apiaries
Ivy Farm
Lidsing Road
Lidsing
Nr Gillingham
Kent ME7 3NL
www.bbha.biz

Brunel Microscopes Ltd
Unit 2
Vincents Road
Bumper's Industrial Estate
Chippenham
Wiltshire SN14 6NQ
www.brunelmicroscopes.co.uk

C. Wynne Jones
Ty Brith
Pentre Celyn
Ruthin LL15 2SR
www.beesupplies.co.uk, or
www.bottlesandjars.co.uk

Caddon Hives
Caddon Lands
Mill Bank Road
Clovenfords
Galashiels
Scotland TD1 3LZ
www.caddon-hives.co.uk

Circomb
29 Glamis Road
Dundee
Scotland DD2 1TS
www.circomb.co.uk

Compak (South) Ltd
3 Ashmead Trading Estate
Keynsham
Bristol BS31 1SX
www.compaksouth.com

E.H. Thorne (Beehives) Ltd
Beehive Business Park
Rand
Nr Wragby
Market Rasen
Lincolnshire LN8 5NJ
www.thorne.co.uk

Fragile Planet
Unit 14
Radfords Field
Maesbury Road
Oswestry SY10 8RA
www.fragile-planet.co.uk

Freeman and Harding Ltd
Unit 18
Bilton Road
Erith
Kent DA8 2AN
www.freemanharding.co.uk

H.S. French Flint Ltd
Unit 4G
The Leathermarket
London SE1 3ER
www.frenchflint.com

KBS
Brede Valley Bee Farm
Cottage Lane
Westfield
Hastings
East Sussex TN35 4RT
www.kemble-bees.com

Maisemore Apiaries
Old Road
Maisemore
Gloucester GL2 8HT
www.honey-online.co.uk,
or www.bees-online.co.uk

Modern Beekeeping
Rooftops
Ebrington Street
Kingsbridge
Devon TQ7 1DE
www.modernbeekeeping.co.uk

National Bee Supplies
Merrivale Road
Exeter Road Industrial Estate
Okehampton
Devon EX20 1UD
www.beekeeping.co.uk

Park Beekeeping Supplies
Unit 17 Blackheath Business Estate
78B Blackheath Hill
London SE10 8BA
www.parkbeekeeping.com

Stamfordham Ltd
Heugh House
Heugh
Newcastle-upon-Tyne NE18 0NH
www.stamfordham.biz

Vita (Europe) Ltd
21/23 Wote Street
Basingstoke
Hampshire RG21 7NE
www.vita-europe.com

GOVERNMENT AGENCIES

England and Wales
The National Bee Unit
The Food and Environment Research Agency
Sand Hutton
York YO41 1LZ
01904 462510
www.nationalbeeunit.com

Scotland
Your local Scottish Government Rural Payments Inspections Directorate (SGRPID) Area Office
08457 741741 or 0131 556 8400
(See also www.scotland.gov.uk/Topics/farmingrural/Agriculture/AOcontacts/contacts)

Northern Ireland
The Department of Agriculture and Rural Development
Dundonald House
Upper Newtownards Road
Belfast BT4 3SB
Northern Ireland
028 9052 4999

Republic of Ireland
Teagasc
Bee Diagnostic Unit
Malahide Road
Dublin 17
Republic of Ireland

Glossary

Abdomen – The third and largest body segment of the bee containing the heart, stomach and intestines. In the worker it also contains the sting and wax glands; in the drone the testes; and in the queen the ovaries and the spermatheca.

Abscond – When a colony abandons its nest completely and flies off to a new location.

Acarine – Disease caused by a mite (*Acarapis woodi*), which infests the bee's tracheae, leading from the first pair of spiracles on the thorax.

AFB – See *American foul brood*.

Afterswarm – The same as a swarm, except that it may contain one or more unmated queens. Also known as a cast.

Alighting board – A strip of wood, usually fixed to the hive stand or as part of the floor, protruding in front of the entrance, on which bees can land before running into the hive.

American foul brood – A notifiable disease caused by a spore-forming bacterium (*Paenibacillus larvae* subspecies *larvae*). Develops in the gut of the larva and kills it after the cell is sealed. Cappings are sunken and perforated. An advisory leaflet on American foul brood is available from the National Bee Unit (www.nationalbeeunit.com).

Amoeba – A protozoan, *Malpighamoeba mellificae*, which affects the bee's equivalent of the kidneys.

Anaphylactic shock – A severe reaction resulting from an acute allergy to bee venom. It may cause death unless immediate medical attention is received.

Antennae – The pair of 'feelers' on the bee's head carrying sensory cells for touch, smell and vibration.

Apiary – The place where one or more hives are kept.

Apiculture – The practice of keeping bees.

Apiguard® – A thymol-based treatment for control of *Varroa destructor*.

Apilife Var® – A treatment for control of *Varroa destructor* based on thymol, menthol, eucalyptus and camphor.

Apistan® – A slow-release polymer strip pyrethroid formulation designed for control of *Varroa destructor*.

Ashforth feeder – A wooden feeder covering the hive cross-section, with one syrup reservoir to which bees gain access from one side.

Bayvarol® – A slow-release polymer strip pyrethroid formulation designed for control of *Varroa destructor*.

BBKA – British Beekeepers' Association.

Beehive – A container for housing honey bees, consisting of a floor, brood box, one or more supers, an inner cover and roof.

Bee space – The space left by bees between the comb and other surfaces in the hive. It's large enough for queen, worker and drone to pass through.

Bee suit – A suit worn by a beekeeper for protection and comfort while opening beehives.

Bee-tight – The situation where the only access to the hive is through the entrance.

Beeswax – A hydrocarbon produced from glands on the underside of the abdomen of the worker bee. Used for comb building and capping cells.

Brace comb – Bridges of wax built between adjacent surfaces in the hive.

Brood – The immature stage of the bee's development. Cells containing eggs and larvae are known as open brood. Sealed cells in which the larvae pupate into adult bees are known as sealed brood.

Brood box – The area in which the queen is confined and the brood is reared.

Brood food glands – See *Hypopharyngeal glands*.

Brood pattern – The pattern of concentric swirls of brood at different stages of development. A good brood pattern has few empty cells, indicating that the queen's brood is largely healthy.

Burr comb – Wax built on a comb or a wooden part in a hive but not connected to any other part.

Canadian escape board – A board for clearing bees from honey supers. Bees pass through narrow gaps but are unable to find a way back in. Available in many different patterns.

Capped brood – See *Sealed brood*.

Capping – The beeswax covering sealing a cell. Cappings over honey consist of wax only. Those over brood include hair and other materials.

Cast – See *Afterswarm*.

Caste – A different form of the same sex. Bees have two female castes – queen and worker.

Castellations – Thin pieces of metal into which slots are cut to take frame lugs. Fastened to opposite inside upper edges of the hive body. Designed to maintain a constant spacing between frames and available with differing numbers of slots.

CCD – See *Colony Collapse Disorder*.

Cells – Small, six-sided wax compartments making up honey comb. Used to store honey and pollen and to rear the juvenile life stages of bees.

Chalk brood – Disease caused by a fungus (*Ascosphaera apis*), which affects sealed brood.

Clearer board – A board covering the hive cross-section, designed to accommodate one or two Porter bee escapes.

Cleansing flight – A flight made by bees that have been confined to the hive for long periods in winter or during bad weather. Bees avoid defecating inside the hive and make a cleansing flight when the weather improves.

Cluster – See *Winter cluster*.

Cold way – Frames arranged at right angles to the entrance of a hive. See also *Warm way*.

Colony – The viable living unit for honey bees, comprising a queen and workers. During the summer male drones are also present.

Comb – The mass of six-sided beeswax cells built by honey bees in which brood is reared and honey and pollen are stored.

Contact feeder – A feeder that gives bees direct contact with the contents. It doesn't cover the whole of the hive surface area and must be surrounded by an 'eke' or empty super so that the roof can be replaced tightly.

Corbicula – See *Pollen baskets*.

Cover board – See *Inner cover*.

Crystallisation – See *Granulation*.

Cut comb – Natural comb or comb built on thin foundation, cut to a size to fit a container for subsequent sale.

Cut comb foundation – Very thin sheets of beeswax foundation as close as practicable to the thickness of the midrib in naturally built comb.

DCA – See *Drone congregation area*.

Dextrose – See *Glucose*.

DN1 – A deep frame with narrow side bars made to fit Modified National brood boxes.

Drawn comb – Cells drawn out by the worker bees to their full depth.

Drifting – The tendency of bees from one colony to accidentally enter another when returning from foraging flights.

Drone – The male bee, whose main function is to fly to a drone congregation area and mate with a virgin queen.

Drone comb – Sections of the comb built for raising drones. The cells are slightly larger than worker cells and have a convex, domed capping when sealed.

Drone congregation area – Place where drones congregate to mate with virgin queens which travel to the same areas.

Drone-laying queen – A queen that lays only unfertilised eggs that develop into drones.

Dummy frame – A slab of wood cut to the same size as a frame to take its place in a hive.

Dysentery – Caused by an excessive amount of water in a bee's body. Afflicted bees defecate inside the hive. Usually caused by prolonged confinement during winter and early spring and consumption of food with a high water content. Often, but not invariably, associated with Nosema.

EFB – See *European foul brood*.

Egg – The first stage of honey bee metamorphosis. Eggs laid by the queen appear as small, thin rods about 1.6mm long, usually placed in the bottom of the cell.

Eke – Four pieces of wood nailed together into a square the same size as the hive. Used to extend the hive when required.

Entrance – The elongated space across the front of a beehive through which bees exit and enter the hive.

Entrance block – A removable block of wood used to reduce the width of the hive entrance.

European foul brood – Disease caused by a bacterium (*Melissococcus plutonius*). Infects the gut of the developing larva and competes for food. Does not kill all affected larvae. A notifiable disease in the UK. An advisory leaflet on European foul brood is available from the National Bee Unit (www.nationalbeeunit.com).

Fermentation – The chemical breakdown of honey, caused by sugar-tolerant yeast and associated with honey having a high moisture content. Used to advantage when making mead.

Fertile queen – A queen, inseminated instrumentally or mated with drones, that can lay fertilised eggs.

Flying bees – Worker bees old enough to go out foraging for nectar and pollen. Foraging generally starts at around three weeks of age.

Following – The annoying habit of some bees to follow and possibly sting another animal coming near to their nest.

Foragers – See *Flying bees*.

Foraging – The act of seeking and collecting nectar, pollen, water and propolis.

Foundation – Beeswax sheets impressed with the shape of cell bases. Available in sizes suitable for worker and drone cells. It can be strengthened with wires or used without.

Frame – Wooden or plastic structure designed to hold comb and enable the beekeeper to inspect and utilise it fully.

Frame runner – A narrow piece of folded metal fastened to opposite inside upper edges of the hive body on which the frames are suspended.

Frame spacers – Plastic or metal spacers which fit over frame lugs and butt up to the spacer on the adjacent frame to ensure constant spacing. Can be narrow or wide.

Fructose (levulose) – The predominant simple sugar found in honey.

Glucose – One of the two principal sugars that constitute honey. The other principal sugar in honey is fructose (levulose).

Glucose oxidase – An enzyme which converts glucose to gluconic acid and hydrogen peroxide during the conversion of nectar to honey.

Granulation – Process that occurs when crystals are formed naturally in honey by the least soluble sugar (dextrose), especially when its temperature falls.

Guard bees – Bees that wait at the hive entrance to guard it from invaders, such as bees from other colonies, wasps, animals or humans. Guard bees give off an alarm pheromone (scent) if the hive is disturbed or threatened and are the first to fly at and attack the invader.

Hefting – The act of slightly lifting a hive from its support to ascertain its weight.

Hive – A man-made structure intended as a home for bees. The best hives allow beekeepers to inspect all aspects of bee life.

Hive stand – A structure that supports the hive and raises it off the ground.

Hive tool – The composite lever/scraper used in the manipulation of a colony.

Hoffman frame – A type of self-spacing frame.

Honey – Concentrated form of nectar that will keep for a long time. The colour and flavour of honey depends on the flowers from which the nectar is gathered.

Honey crop – An organ in the bee's thorax used for carrying nectar, honey or water.

Honey extractor – Machine that allows honey to be extracted from combs so that they can be reused.

Honey flow – A heightened influx of nectar into the hive brought about by favourable weather conditions and the availability of suitable flowers.

Honey ripener – See *Settling tank*.

Honeydew – The product of sap-sucking bugs such as aphids. Collected by bees when it's diluted by dew early in the day.

House bee – A young worker that stays in the hive and performs tasks such as feeding young larvae, cleaning cells and receiving and storing nectar and pollen from foragers.

Hypopharyngeal glands – Glands in the head of the worker bee which produce brood food and royal jelly to be fed to the developing larvae. As the worker bee matures, the glands decrease in size and switch to production of the enzymes invertase and glucose oxidase, which help to convert nectar to honey.

Inner cover – A board that's placed over the frames just beneath the roof. Also called a 'quilt' or cover board.

Integrated Pest Management – The use of substances and specific colony manipulations to reduce the population of *Varroa destructor*.

Invert sugar syrup – A liquid sugar syrup formed by the chemical breakdown of sucrose resulting in an equal mixture of glucose (dextrose) and fructose (levulose).

Invertase – An enzyme that converts sucrose to glucose and fructose during the conversion of nectar to honey.

IPM – See *Integrated Pest Management*.

June gap – A period during June when availability of forage is seriously reduced. Colonies could starve unless the beekeeper checks their stores and feeds them if necessary.

Larva – The second stage of bee metamorphosis. The larva hatches from the egg, develops into a pupa and changes into an adult.

Laying worker – A worker that lays unfertilised but fertile eggs, producing only drones. Occurs if a colony becomes queenless and is not able to raise a new queen.

Levulose – See *Fructose*.

Mandibular glands – Glands in the head of the worker bee which produce brood food and royal jelly, which are fed to the developing larvae.

Mating flight – Flight taken by a virgin queen, which mates in the air with several drones.

Mead – Alcoholic wine-like drink made from honey and water.

Melomel – Type of mead made using fruit juice and honey.

Metal ends – See *Frame spacers*.

Metal runner – See *Frame runner*.

Migratory beekeeping – The moving of colonies of bees from one locality to another during a single season to take advantage of two or more honey flows.

Miller feeder – A wooden or plastic feeder of the same cross-sectional size as the hive. There are two syrup reservoirs, to which bees gain access from a central slot.

Modified National – The commonest hive in use in the United Kingdom. A single-walled hive.

Moult – The shedding of its skin by a larva to make room for new growth.

Mouseguard – A metal strip or similar containing holes that allow bees in and out of the hive but prevent mice from gaining access.

NBU – National Bee Unit.

Nectar – The sugary secretion of plants produced to attract insects for the purpose of pollination.

Nectar guides – Marks on flowers believed to direct insects to nectar sources. They may be visible to the human eye or may reflect ultraviolet and hence be visible only to bees.

Nosema – Disease caused by specialised microsporidian fungal pathogens (*Nosema apis* and *Nosema ceranae*). These infect the gut of the bee and shorten its life by preventing it from properly digesting its food.

Nucleus – A small colony, usually on three, four or five frames. Used primarily for starting new colonies, or for rearing or storing queens. Also known as a 'nuc'.

Nucleus hive – A small hive designed to contain three, four or five frames only.

Nurse bees – Young worker bees, three to ten days old, which feed and take care of developing brood.

Orientation flight – A short flight taken by a young worker in front of or near the hive prior to when it starts foraging, in order to establish the position of the hive. Also referred to as a 'play flight'.

Out-apiary – An apiary established away from the beekeeper's home.

Pheromone – A substance produced by one living thing that affects the behaviour of other members of the same species. Pheromones produced by the queen help the colony to function properly.

Plastic ends – See *Frame spacer*.

Play cells – See *Queen cell cups*.

Play flight – See *Orientation flight*.

Pollen – The part of the plant carrying the male contribution to the production of future generations.

Pollen basket – The corbicula, a segment on the hind pair of legs in a worker bee specifically designed for carrying pollen. Also used to bring propolis back to the hive.

Pollen load – Pellets of pollen carried by a foraging worker bee in the pollen baskets (corbiculae) on its hind pair of legs.

Pollination – The transfer of pollen from the anthers to the stigma of flowers.

Porter bee escape – A device used for clearing bees from supers. Two spring valves allow bees to pass through one way but not to return.

Prime swarm – The first swarm to leave the colony, usually containing the old queen.

Proboscis – The mouthparts of the bee that form the sucking tube or tongue. Used for sucking up liquid food (nectar or sugar syrup) or water.

Propolis – A resinous material collected by bees from the opening buds of various trees, such as poplars.

Pupa – The third stage in the metamorphosis of the honey bee, during which the organs of the larva are replaced by those that it will use as an adult. Takes place in a sealed cell.

Queen – One of the two variants or castes of female bees. Larger and longer than the worker bee.

Queen cell cups – The base of a queen cell into which the queen will lay an egg designed to develop into a new queen.

Queen cell – An elongated brood cell hanging vertically on the face of the comb, in which a queen is reared.

Queen excluder – A device with slots or spaced wires that allows workers to pass through but prevents the passage of the queen and drones.

Queen substance – Complex pheromones produced by the queen. Transmitted throughout the colony by the exchange of food between workers to alert other workers of the queen's presence. Its presence stops worker bees rearing more queens and/or inhibits them from laying eggs.

Queenless – When a colony has no queen. If bees have access to worker eggs or very young larvae, they're able to rear a replacement queen.

Queenright – When a colony has a living, laying queen.

Quilt – See *Inner cover*.

RBI – Regional Bee Inspector.

Retinue – Worker bees who attend the queen and care for her needs within the hive.

Ripe queen cell – A queen cell that is near to hatching. Bees remove wax from the tip, exposing the brown parchment-like cocoon.

Robbing – When wasps, or bees from other colonies, try to steal honey from a hive.

Round dance – A communication dance used by bees to indicate a food source up to 100m (300ft) from the hive.

Royal jelly – A highly nutritious glandular secretion of young bees, used to feed the queen, young brood and larvae being reared as new queens.

Sac brood – A viral disease that prevents the final larval moult. The larva dies in its larval skin, which is removed from the cell.

SBI – Seasonal Bee Inspector.

Scout bees – Worker bees that search for new sources of nectar, pollen, water and propolis. If a colony is preparing to swarm, scout bees will search for a suitable location for the colony's new home.

Sealed brood – The pupal stage in a bee's development during which it changes into an adult.

Sections – Honey comb built into special bass wood frames. Generally sold complete. Also available in circular plastic form.

Self-spacing frame – A frame in which the upper part of the side bar is extended to touch that of the adjacent frames, maintaining a constant distance between them.

Settling tank – A holding tank for honey that allows air to rise to the surface before bottling using the tap at the base.

SHB – See *Small hive beetle*.

Shook swarm – Bees shaken, together with their queen, from one hive into another. Used to control swarming or diseases such as *Varroa destructor* or European foul brood.

Skep – A beehive constructed from straw that does not contain movable frames. No longer in general use as a permanent home for a colony. Now often used for collecting swarms.

Small hive beetle – A small beetle (*Aethina tumida*) about one-third the size of a worker bee. Dark red, brown or black. Has distinctive clubbed antennae. Both larvae and adults eat honey and pollen. Will spoil honey in the comb. A growing pest in the USA although not yet thought to be present in the UK, where it is a notifiable disease. An advisory leaflet on small hive beetle is available from the National Bee Unit (www.nationalbeeunit.com).

Smoke – The product of burning suitable materials in a smoker. The best smoke for working with bees comes from organic materials such as rotten wood, shavings, etc.

Smoker – Device that delivers smoke in a precise manner.

SN1 – A shallow frame with narrow side bars made to fit Modified National super boxes.

Spermatheca – A special organ in the queen's abdomen in which she stores sperm received from drones during mating.

Spiracles – Apertures found on the sides of the thorax and abdomen that are the openings of the breathing tubes or tracheae.

Sting – The defensive mechanism at the end of the abdomen used by worker bees to deter predators. The queen will use her sting to kill rival queens, usually when several are hatching or due to hatch during the swarming process.

Stores – The weight of honey collected by bees, especially the reserves needed for winter.

Super(s) – The box(es) placed on top of the brood chamber to increase the space available for honey storage.

Supersedure – The raising of a replacement queen when the old queen is still present in the colony. The two may live together in harmony until the old queen dies.

Sugar syrup – A solution of sugar and water used to feed bees.

Swarm – A mass of bees not in a hive. The bees may be wanting to establish a new colony or be absconding from a bad environment. It should contain a mated queen.

Swarm cells – Queen cells, often but not always found on the lower edges of the combs before swarming.

Swarm control – Methods used to stop a swarm from leaving the hive.

Swarm prevention – Methods used to prevent the physical conditions arising that stimulate a colony to prepare to swarm.

Thin foundation – A sheet of foundation which is thinner than that used for brood rearing. Used for comb honey production.

Thorax – The second, central part of the bee's body. It contains the flight muscles and has the legs and wings attached.

Tropilaelaps – *Tropilaelaps clareae* and *Tropilaelaps koenigerum* are serious parasitic mites which affect both developing brood and adult honey bees. They're not currently known to be in the UK but are a statutory notifiable pest of honey bees. An advisory leaflet on *Tropilaelaps* is available from the National Bee Unit (www.nationalbeeunit.com).

Uncapping knife – A knife used to remove the cappings from combs of sealed honey prior to extraction.

Uniting – The act of combining two or more colonies to form a larger colony. Colonies are usually united if one is weak or has lost its queen.

Varroa destructor – A mite that breeds in sealed brood cells, feeding on the larval blood. If it doesn't kill the developing larva, it can trigger viruses that lead to serious deformations such as shrivelled wings.

Veil – The bee-proof see-through hood worn by beekeepers to protect them against stings.

Venom – Poison secreted by special glands attached to the bee's sting.

Venom allergy – A condition in which a person, when stung, may experience a variety of symptoms ranging from a mild rash or itchiness to anaphylactic shock.

Virgin queen – A young, unmated queen.

Waggle dance – The most common communication dance used by bees to indicate a food source over 100m (300ft) from the hive.

Warm way – Frames arranged parallel to the entrance of a hive. See also *Cold way*.

Wax glands – Four pairs of glands on the underside of the last four visible abdominal segments of the worker bee that secrete small particles of beeswax.

Wax moth (greater) – The greater wax moth (*Galleria mellonella*) primarily infests stored equipment but will invade colonies where the worker bee population has been weakened. The larvae chew into woodwork to make depressions in which to pupate.

Wax moth (lesser) – The lesser wax moth (*Achroia grisella*) has the same type of scavenging habits as the greater wax moth but causes less damage. The adults are similar to clothes moths and are characterised by a yellow head.

WBC hive – A double-walled hive designed by William Broughton Carr.

Windbreak – A barrier, such as a thick hedge, to break the force of the wind blowing on to hives in an apiary.

Winter cluster – The roughly spherical mass adopted by bees as a means to survive the winter.

Worker – The commonest bee in the colony, specialised to undertake the tasks required for the continuation of the colony such as feeding young larvae and foraging for nectar and pollen.

Worker comb – Sections of the comb built for raising worker bees. When sealed, the cappings are flat. Also used for storing honey and pollen.

Index